An Introduction to Passive Radar

For a complete listing of titles in the
Artech House Radar Series,
turn to the back of this book.

An Introduction to Passive Radar

Hugh D. Griffiths

Christopher J. Baker

ARTECH HOUSE

BOSTON | LONDON
artechhouse.com

Library of Congress Cataloging-in-Publication Data

A catalog record for this book is available from the U.S. Library of Congress.

British Library Cataloguing in Publication Data

A catalogue record for this book is available from the British Library.

Cover design by John Gomes

ISBN 13: 978-1-63081-036-8

685 Canton Street
Norwood, MA 02062

10 9 8 7 6 5 4 3 2 1

Contents

Foreword

A passive radar is a fascinating device that does not transmit any electromagnetic energy when sensing. It instead listens to existing illuminations and senses tiny disturbances in the electromagnetic field caused by moving objects in order to detect and track them, and even provide imaging. The idea is not new, as it was in 1934 that Robert Watson Watt made his famous Daventry experiment, proving that through the use of short-wave commercial transmitters, it was possible to detect a bomber at long distances, up to 8 miles in his case. A few years later, German engineers built the first operational passive radar, which they named Klein Heidelberg. The device used British Chain Home radars as illuminators and was able to detect British bombers as they flew on their missions to destroy military targets during World War II.

Despite being widely known, the idea of passive radar was discarded for several decades as analog signal processing was not sufficiently advanced for the effective processing of passive radar signals.

The rapid development of digital signal processing methods in the 1970s opened new possibilities regarding the implementation of the concept, and research on passive radar began soon afterwards in the United Kingdom, Germany, Italy, France, Poland, China, Iran, Russia, and several other countries. The U.S.-based company Lockheed Martin went on to build a series of demonstrators of the new technology in the 1990s that they called Silent Sentry I, II, and III.

Research work on passive radar was initiated by Professor Griffiths at University College London in 1982. He soon extended his research to the exploitation of commercial transmitters of opportunity and, in 1986, published the first paper on passive radar using analog ultrahigh frequency television (UHF TV) transmitters as illuminators.

In 1989, he started working with Professor Baker on exploiting a wide range of transmitters of opportunity for passive sensing, including satellite TV, and together they continue to conduct research in this field to this day. The two have published more than 100 scientific papers on this subject, becoming scientific leaders in the area of passive radar.

This book is the first on the market to present passive radar technology to the general public. As stated in the title, the authors are looking to provide the reader with an introduction to passive radar. In a very simple and clear way, they present how passive radar works and how it differs from the active type and show both the benefits and drawbacks of this novel technology. This book is addressed to all who intend to understand this new technology without going deeply into mathematical descriptions of the problems involved. This book also shows the history of passive radar development, principles of operation, the property of different illuminators of opportunity, and signal and data processing fundamentals including correlation reception, direct signal cancelation, bistatic detection, target localization, and tracking. Practical aspects are also discussed, and selected demonstrators and final products are presented.

Although fundamental mathematical formulas related to passive radar phenomena and processing are presented, the reader can easily skip them and still fully grasp the basic phenomena and principles essential to the understanding of the technology. Advanced readers can analyze the equations and find references to a number of scientific publications where all the mathematical details can be found.

The passive radar market is currently an emerging one, and very few products have been made available at the present time. However, a lot of industrial competitors have started new programs aiming towards the development of new passive radars both for the civil and military markets, worth more than $10 billion in U.S. currency.

The spectrum is becoming more congested and valuable as time goes on. Communication companies are trying to get more spectra for their purposes, and it is more difficult now to get a license for emitting electromagnetic power for radar and remote sensing purposes. The passive radar is a "green" solution to this problem: there is no need to pay for transiting energy for spectrum allocation. Only sensitive receivers, digitizers, and fast signal processing devices are needed to construct radars that could be used for air traffic control, border protection, military purposes, and even for bird migration analyses.

In this book you will find the knowledge needed to gain an understanding of this fascinating technology, and hopefully you will come away with a fresh perspective and deeper insight into the passive radar's inner workings after reading it.

Professor Krzysztof Kulpa
Warsaw University of Technology
February 2017

Preface

The motivation for writing this book came from an approach by Dr. Joe Guerci, the Artech Radar Series editor. He pointed out that over the years we had accumulated a large volume of research results and lecture and tutorial material, and that it should not be difficult to assemble it into a book. Of course, it was not quite as simple as that, and the project has required many hours of organization of material, writing, rewriting, and checking.

The historical introduction in Chapter 1 shows that passive radar has been around for a long time—arguably for more than 90 years. Because passive radar systems do not need dedicated high-power transmitters and are generally simple and low-cost, the subject has been very suitable for university research, and numerous university groups worldwide have worked on passive radar and published their results at conferences and in research journals. But there was always a sense that passive radar was "almost as good as a real radar," and seemed to be an answer looking for the right problem. In the past 5 years this situation has certainly changed, and passive radar has come of age. The widespread use of digital transmissions for broadcast and for communications has provided waveforms that are much more suitable for radar use. The spectrum congestion problem has provided the impetus to seek sensing techniques that exploit existing transmissions rather than adding to the congestion. Commercial companies in several countries have invested resources to develop practical passive radar systems that offer much better performance and reliability than the experimental systems of a decade ago. In the next decade we can expect further advances, and the final chapter identifies a number of passive radar applications that seem particularly promising.

This is not a detailed mathematical treatise. It has been our intention to produce a book that is accessible to anyone with a basic understanding of radar. Our emphasis has been on presenting the essential principles, processing techniques, and practical results. We have tried to provide comprehensive lists of references so that the reader can access the original publications in full detail having been suitably equipped with the basics. This book draws on previously published books and research, most notably Nick Willis's book on bistatic radar, which provides an excellent introductory text to the more general topic of bistatic radar, and references to this occur in a number of the chapters.

We have both spent a substantial part of our careers working on passive radar research, and we have been lucky to have worked with many talented radar engineers, all of whom have made their own particular contributions to progress. There is very much the sense of a global community, sharing ideas and results, and building on each other's work. It is a cast of thousands that has provided us with much inspiration as well as innumerable discussions that have helped to shape this book. While it is impossible to acknowledge everyone, we mention particularly our close colleagues: Matt Ritchie, Alessio Balleri, Graeme Smith, Andy Stove, Simon Watts, and Landon Garry.

We also express our thanks to Dr. Joe Guerci for suggesting the project in the first place, and to Molly Klemarczyk and Aileen Storry of Artech for keeping us on track. We thank the various authors, organizations, and publishers who have given permission for the use of figures, and we are grateful to colleagues who have read early drafts, corrected errors, or suggested better ways of expressing ourselves. But responsibility for any remaining errors rests with us.

But most of all we thank our wives, Morag and Janet, for their encouragement and their continuing forbearance.

CHAPTER

1

Contents

Introduction

1.1 Terminology

Passive radar may be defined as a set of radar techniques that exploit existing signals, such as broadcast, communications, or radionavigation emissions as their transmitting source. This contrasts with conventional monostatic radars, which use their own dedicated transmitter and a single antenna for both transmitting and receiving, and where the form of the signal (usually pulsed) is optimized solely for the radar function.

Although the term *passive radar* has been widely adopted, other terms have been suggested and used. *Passive coherent location* (PCL) and *passive covert radar* (PCR) have been used, particularly by military users. *Passive bistatic radar* (PBR) is widely used and emphasizes the bistatic configuration with physical separation of transmitter and receiver, which in practice dictates many of the properties of this type of radar. The term *hitchhiking* is used when the transmitter source is an existing monostatic radar. Yet further terms that have been used include *broadcast radar, noncooperative radar, parasitic radar,* and *symbiotic radar* [1]. However, for the purposes of this book, we have opted to keep it simple, so *passive radar* it is.

The transmit sources used in passive radar are often referred to as *illuminators of opportunity,* which means that their signals are optimized for some purpose other than radar. A further distinction has been made between *cooperative sources* and *noncooperative sources.* In practice, these definitions are rather crude and there is actually a range of situations. At one extreme, the waveform and coverage of the illuminator might be completely cooperative and under the control of the passive radar designer and could even be dynamically varied in real time. At the other extreme, an illuminator might be designed so as to be as difficult as possible to exploit as a passive radar signal. Clearly, there are a number of cases in between these extremes.

One variation that is of particular current interest is *commensal radar,* in which a broadcast or communications signal is designed so that it not only fulfills its primary purpose, but is also in some sense optimized as a radar signal as well. (The term commensal radar has also been used by researchers in South Africa as a synonym for passive radar, especially with noncooperative illuminators.) This takes advantage of the sophisticated digital waveform coding techniques that modern digital signal processing now allows, also known as waveform diversity. Commensal radar may be useful as a way of addressing the spectrum congestion problem and is described in more detail in Chapter 8.

A close cousin of passive radar is passive emitter tracking (PET), which locates and tracks targets (usually aircraft) on the basis of signals emitted by the targets and received at multiple receivers.

Passive radar has a number of potential attractions:

- Broadcast and communications transmitters tend to be sited on high locations and hence achieve broad coverage.
- Because the system makes use of existing transmitters, the cost of a passive radar is likely to be much lower than a conventional radar.
- Similarly, there are no licensing issues.
- It allows the use of frequency bands [particularly very high frequency (VHF) and ultrahigh frequency (UHF)] that are not normally available for radar purposes. Such frequencies may be beneficial in

detecting stealthy targets, as the wavelength is of the same order as the physical dimensions of the target, and forward scatter gives a relatively broad angular scatter.

- Because the receiver emits no signal of its own, and as long as the receive antenna is inconspicuous, the passive radar receiver may be undetectable and hence completely covert.
- It is difficult to deploy countermeasures against passive radar. Any jamming will have to be spread over a range of directions, diluting its effectiveness.
- Passive radar does not require any additional spectrum. For this reason, it has been termed "green radar."
- There is an enormous range of transmissions that may be used. In practice, almost any emission can be used as the basis of a passive radar.

However, there are also some significant disadvantages:

- The waveforms of such transmissions are not optimized for radar purposes, so care has to be used to select the right waveforms and to process them in the optimum way.
- In many cases, the transmit source is not under the control of the passive radar.
- For analog signals, the ambiguity function (resolution in range and in Doppler) depends on the instantaneous modulation, and some kinds of modulation are better than others. Digital modulation does not suffer from these problems, so is likely to be preferred.
- The waveforms are usually continuous (i.e., a duty cycle of 100%), so significant processing has to be used to suppress the direct signal and multipath in order to detect weak target echoes.
- In common with all bistatic radars, the resolution in range and Doppler is poor for targets on or close to the baseline between transmitter and receiver.

These points are discussed in greater detail in the chapters of this book.

1.2 History

As is often the case, the history of the subject goes back further than might be expected. Some of the very first radar experiments in the United States, by A. Taylor and L.C. Young of the Naval Research Laboratory in 1922, were bistatic [2]. Arguably, the first documented use of a broadcast transmitter for radar purposes was in 1924, when Appleton and Barnett [3] used a broadcast radio transmitter located at Bournemouth on the south coast of England, at a frequency of approximately 770 kHz, and a receiver located at Oxford at a distance of approximately 150 km, to measure the height of the Heaviside layer (ionosphere). Here, the signal traveled by two paths: the direct ground wave path, and the path reflected from the ionosphere (Figure 1.1). The two signals combined with a particular phase relationship given by the difference in path length. In a letter to a colleague, Appleton explained that at this range the two signals would be of approximately the same amplitude, and in practice this was indeed found to be the case [4]. The frequency of the transmitter was swept over a bandwidth of about 20 kHz with a sweep period of about 10 seconds. In this way the phase relationship changed, tracing out a set of maxima and minima on a crystal rectifier and galvanometer used to detect the signals, with a separation proportional to the path difference. This can also be considered therefore as the first frequency modulation (FM) radar [5].

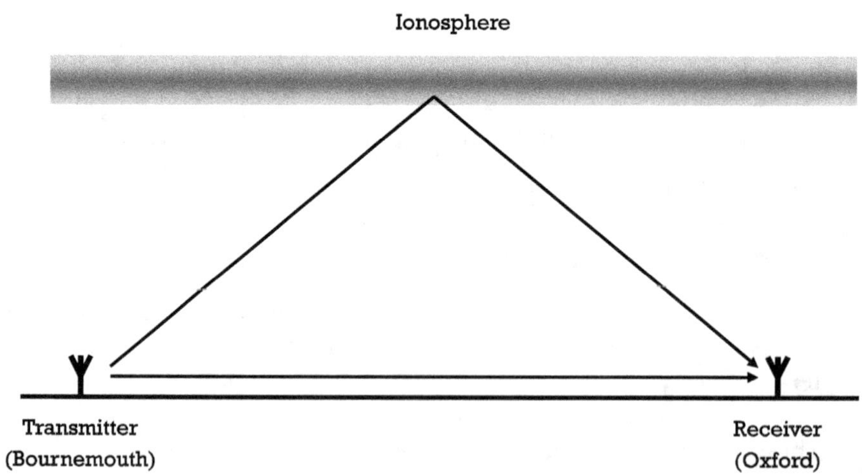

Figure 1.1 Appleton and Barnett's 1924 experiment.

Another milestone in radar history was the celebrated Daventry experiment [5, 6] on February 26, 1935. Robert Watson Watt (whose background was also in ionospheric science) and his assistant Arnold Wilkins used a British Broadcasting Corporation (BBC) broadcast transmitter at a frequency of around 6 MHz to detect an aircraft target (a Handley Page Heyford bomber) at a range of 8 miles. More importantly, they were able to demonstrate this result to a senior civil servant (A. P. Rowe), convincing the British Air Ministry to fund a program of development that led to the British *Chain Home* air defense radar system, just in time for the outbreak of World War II [6].

As early as 1938, a publication in an American scientific magazine reported the "ghosting" effect on a television screen caused by reflections from an aircraft target [7]. Here, the aircraft echo represents a slightly-delayed version of the direct signal, so it appears as a ghost image on the screen.

We should also mention the German World War II bistatic radar system *Klein Heidelberg* (Figures 1.2 and 1.3), which used the British Chain Home radars as a hitchhiking illumination source, and was the first example of an operational passive radar [8, 9]. The first Klein Heidelberg became operational in 1943, with five further stations being built, and the Allies did not find out about it until October 1944 [10]. It proved

Figure 1.2 (a) The antenna and bunker of the Klein Heidelberg at Oostvoorne, The Netherlands, in 1947. (b) The expanded view shows the dipole elements and the wire reflecting mesh. (Photograph courtesy of Jeroen Rijpsma.)

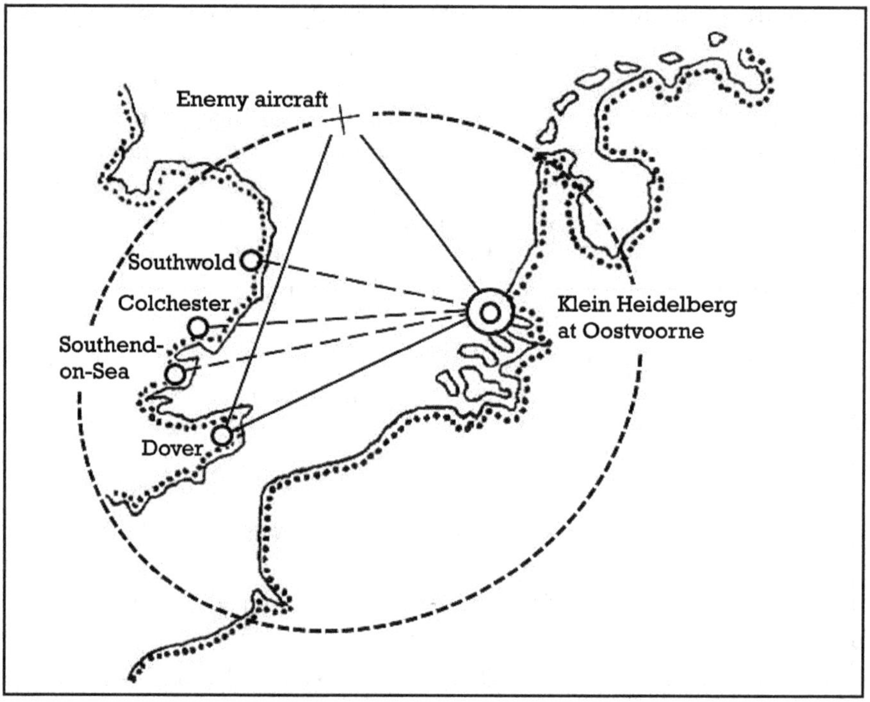

Figure 1.3 Principle of the German Klein Heidelberg bistatic radar system. (Adapted from: [11].)

valuable in German air defense when conventional early warning radar was affected by jamming and other countermeasures, but came too late to have any decisive effect in World War II. An important feature of the Chain Home radars that made this possible is that their transmitters had a broad beam floodlight illumination.

After the end of World War II, interest in bistatic radar waned, because the additional complication of bistatic operation did not provide any significant advantage or capability except in some rather specific applications. Nickolas Willis, who wrote and coedited two of the classical books on bistatic radar [1, 12] and the bistatic radar chapter in Skolnik's *Radar Handbook* [13], observed that bistatic radar has undergone three resurgences. The first resurgence saw deployment in the United States of the A\N/FPS-23 Fluttar in 1957 as an early warning, bistatic fence to detect aircraft in the DEW Line, the 440-L OTH forward scatter system in 1967 to detect ballistic missile launches, and three multistatic systems: the VHF passive ranging Doppler and the microwave Doppler test-range instrumentation radars and the massive SPASUR satellite tracking radar spread across the geographical United States, which after 50 years of continuous use was recently retired. Perhaps the most significant—and continuing—bistatic contribution is air defense missile guidance, where a target-tracking radar illuminates the target and the missile carries the receiver, a configuration more commonly called semiactive homing. Furthermore, target glint, a significant contribution to miss distance, can be greatly reduced by opening up the bistatic angle between the illuminator, target, and missile.

The second resurgence saw deployment of inexpensive, piggyback bistatic radars to measure moon and planetary surfaces starting in 1967 and continuing for nearly 40 years. A principal activity during this time was the development and testing of many bistatic radar configurations to counter the retro-jamming and antiradiation missile (ARM) threats to monostatic radars both in the United Kingdom and the United States. While many of the tests were successful, none were deployed because other solutions were simpler and cheaper (e.g., decoys, GPS to reduce site location errors, plus fast, dedicated communications links from standoff monostatic radars). One multistatic radar, the multistatic measurement system, was installed at the U.S. Kwajalein Missile Range in 1980 to reduce monostatic cross-range measurement errors. It operated successfully for 13 years and was then removed.

The third resurgence saw further research, development, and testing of bistatic technology, including autofocus algorithms for improving synthetic aperture cross-range measurement accuracy, space-time adaptive processing to improve clutter-limited detection of moving targets, exploiting commercial broadcast transmitters for inexpensive, covert air defense surveillance, and exploiting monostatic weather radars to generate full vector wind fields around airports. Two bistatic systems were deployed in the mid-1990s: the University of Washington developed Manastash Ridge Radar, which uses FM broadcast transmitters to study turbulence in the ionosphere, and the Russian Struna-1 forward-scatter fence to detect low-flying aircraft and provide limited state estimates of nonmaneuvering targets.

These results augur well for further practical bistatic radar systems, both because there continue to be applications in which the performance advantage of bistatic radar is worthwhile, and because the technology and (particularly) processing power now allow these advantages to be realized. Passive radar, inherently bistatic, is one of the main drivers behind the third resurgence.

Publications describing work in the 1950s and 1960s are very sparse, mostly because the work itself was limited, but also because virtually all of it was classified. We can note a paper published by Rittenbach and Fishbein in 1960 [14], which is a description of a concept using a transmitter carried by a geostationary satellite, radiating a randomly-modulated continuous-wave (CW) signal at a power level of 100W. The system was designed to detect ground vehicle targets and the receiver used two antennas, one pointed directly at the satellite to receive the direct signal and the other pointed to receive the target echoes. The processing consisted of a cross-correlation operation between the target echoes and the direct signal. The publication does not give any indication of practical trials or results.

Lyon, in [15] also reported a U.S. Over-The Horizon (OTH) high-frequency (HF) passive bistatic program *Sugar Tree* that was deployed in the 1960s, to detect Soviet missile launches. The transmitter was located close to the launch site, so the direct signal and target echo would both propagate to a remote receiver by sky-wave. Information on this system has only been declassified in recent years [16, 17].

Work at University College London (UCL) in the early 1980s led to the demonstration of a bistatic radar system hitchhiking off an Air Traffic Control radar located at Heathrow Airport [18]. The frequency of this radar was around 600 MHz. The work demonstrated real-time synchronization, coherent moving target indication (MTI), and pulse-chasing using a digital beamforming array. It was then reasoned that UHF television transmissions should also be usable as bistatic radar illuminators, because they are high-power, relatively broadband (~6 MHz), and at a similar frequency to the Air Traffic Control radar. The work provided some of the first experimental demonstration of the concepts of passive radar [19], but also pointed out some of the difficulties of using signals with 100% duty cycle and strong periodic features associated with the line scan rate of the signal. Interestingly, in those days television stations did not transmit 24 hours per day, and prior to startup each day they would broadcast a test card picture to allow television receivers to be correctly aligned. In the bistatic radar experiments, it was possible to arrange for the BBC to radiate special test cards providing waveforms with favorable radar properties.

Some important publications from the following decade can be noted. At the Air Force Research Laboratory in Rome, New York, Ogrodnik [20] demonstrated a low-cost man-portable passive radar system for air surveillance. In 1999, Howland [21] published work demonstrating the use of television signals to detect and track aircraft targets. His system overcame the limitations of the analog television waveform by using just the vision carrier, and measuring Doppler shift and angle of arrival (but not range) of the radar echoes. The Doppler and angle of arrival information was input to an extended Kalman filter tracking process, and the system was able to demonstrate tracking of civilian aircraft over a wide area of the southeast United Kingdom. A second publication, in 2005, demonstrated the use of a single VHF FM transmission and a single receiver to detect and track aircraft targets, out to ranges beyond 100 km [22].

Since then, work on passive radar has grown steadily. It has been a particularly suitable subject for university research groups, because the receive hardware is relatively low-cost and there are no issues with licensing or with high-power transmitters. Many of these experiments have exploited FM radio transmissions, although HF broadcast, digital radio and television, cellphone base stations, WiFi and WiMAX, and various kinds of satellite signals have also been used. A significant development has been the introduction of digital modulation formats. Analog television transmissions were turned off in the United States in 2009. France discontinued all analog services on November 29, 2011, followed by Japan on March 31, 2012; and many other countries have been set to follow. It was announced that FM radio transmissions in Norway would be turned off as of January 2017.

The Manastash Ridge Radar (MRR) was a passive radar using a single FM radio transmitter, built and used at the University of Washington in Seattle for studies of ionospheric physics [23, 24]. The use of passive radar, particularly of an FM radio illuminator source, was regarded as especially suitable, as the receiving system was low-cost and the frequency (~100 MHz) was ideal for this application. The problem of suppression of the direct signal at the receiver was solved by locating the receiver the other side of a mountain range. The system gave continuous monitoring

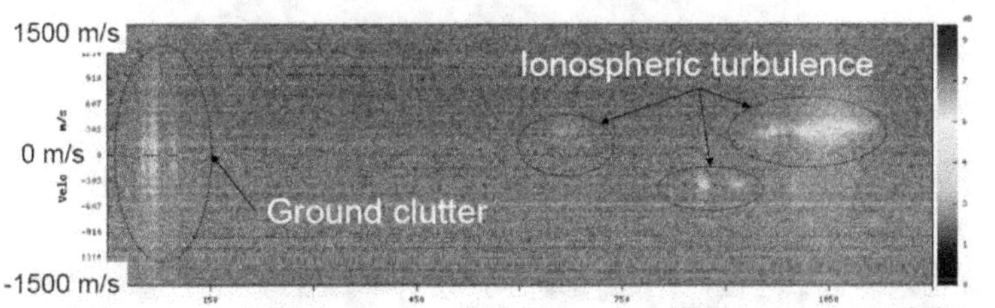

Figure 1.4 Example of the output from the Manastash Ridge Radar, presented in the form of a Range-Doppler plot. (Courtesy of John Sahr.)

of the ionosphere at ranges out to 1,200 km. This system illustrates an attractive set of applications of passive radar as a low-cost approach to remote sensing.

In the late 1990s and early 2000s the U.S. company Lockheed Martin developed a passive radar system called *Silent Sentry* [25]. This uses FM radio illuminators and was able to demonstrate real-time tracking of multiple aircraft targets, as well as real-time detection of rocket launches from Cape Canaveral. More recently, a number of other companies have developed and demonstrated passive radar systems of their own, including *Homeland Alerter* from the THALES Company in France, *AULOS* from the LEONARDO company in Italy, and a multiband FM/DAB/DVT system from Airbus Defence and Space (formerly Cassidian) (see Figure 1.5).

These recent developments mark a change in the maturity of the subject. They exploit multiple kinds of transmissions and achieve much greater coverage and reliability than previous systems. This means that such systems are now seriously being considered for applications such as air traffic control, gap filling in the coverage of conventional radars, and as a potential solution to the spectrum congestion problem [26].

Research on passive radar has benefitted from a number of NATO Task Groups on different aspects of the subject, as well as some substantial projects within the scientific program of the European Union. This has in

Figure 1.5 The antenna of the passive radar demonstrator produced by Airbus Defence and Space.

part accounted for the high degree of interest and activity in European countries.

1.3 Approach and Scope

The aim of this book is to provide an up-to-date introduction of the subject of passive radar. The treatment is intended to be appropriate for a practitioner in industry or a graduate student, and emphasis is placed on practical systems and practical processing techniques.

The structure of the rest of this book is as follows. Chapter 2 covers the properties of bistatic radar. Many of these are a consequence of the bistatic geometry. The chapter also derives the bistatic radar equation and treats the bistatic radar cross section of targets and clutter.

Chapter 3 covers the very important subject of properties of passive radar emitters, and a wide range of these is described. It is found that analog waveforms may have time-varying ambiguity functions which depend on the nature of the program content, while digital modulation formats are much more noise-like and hence better as radar signals. Also important, especially for air target detection and tracking, is the vertical-plane coverage of illuminators.

Chapter 4 treats the subject of suppression of direct signal at the passive radar receiver. The level of such signals may be as much as 100 dB or even more above noise level, so the dynamic range requirements on the passive radar receiver are stringent.

All of these considerations are brought together in Chapter 5, which shows how the performance of passive radar systems may be predicted in a realistic manner by understanding the correct values to insert into the basic radar equation, and the effects of any assumptions and approximations.

Chapter 6 covers the subjects of detection and target tracking in passive radar. The information provided by each transmitter-receiver pair may be one or more of bistatic range, Doppler shift, and direction of arrival. Combining these to provide reliable tracks of multiple targets is not necessarily easy.

Chapter 7 gives examples of a range of practical passive radar systems and their results, arranged following the types of emitters considered in Chapter 3.

Finally, Chapter 8 discusses future developments and applications. Because passive radar is something of a hot topic, the future seems set to be exciting. Indeed, a recent defense business Web site has predicted that the military and civil aviation market for passive radar for the decade 2013–2023 is likely to be worth more than $10 billion in U.S. currency [27].

References

[1] Willis, N. J., and H. D. Griffiths, *Advances in Bistatic Radar*, Raleigh NC: SciTech Publishing, 2007, pp. 78–79.

[2] Glaser, J. I., "Fifty Years of Bistatic and Multistatic Radar," *IEE Proc.*, Pt. F, Vol. 133, No. 7, December 1986, pp. 596–603.

[3] Appleton, E. V., and M. A. F. Barnett, "On Some Direct Evidence for Downward Atmospheric Reflection of Electric Rays," *Proc. Roy. Soc.*, Vol. 109, December 1925, pp. 261–641.

[4] Letter from Edward Appleton to Balth van der Pol, January 2, 1925 (transcribed by B.A. Austin).

[5] Griffiths, H. D., "Early History of Bistatic Radar," *EuRAD Conference 2016*, London, October 6–7, 2016.

[6] Watson-Watt, R. A., *Three Steps to Victory*, Chapter 20, London, U.K.: Odhams Press, 1957, pp. 107–117.

[7] *Science News Letter*, April 23, 1938.

[8] Griffiths, H.D. and Willis, N.J. "Klein Heidelberg: The First Modern Bistatic Radar System," *IEEE Trans. on Aerospace and Electronic Systems*, Vol. 46, No. 4, October 2010, pp. 1571–1588.

[9] Griffiths, H. D., "Klein Heidelberg: New Information and Insight," *IEEE Radar Conference 2015*, Johannesburg, October 2015.

[10] *Air Scientific Intelligence Interim Report, Heidelberg*, A.D.I. (Science), IIE/79/22, 24 November 1944, Public Records Office, Kew, London (AIR 40/3036).

[11] Hoffmann, K. -O., *Ln-Die Geschichte der Luftnachrichtentruppe, Band I/II*, Neckargemünd, 1965 (in German).

[12] Willis, N. J., *Bistatic Radar*, 2nd ed., Silver Spring, MD: Technology Service Corp., 1995, corrected and republished by SciTech Publishing, Raleigh NC, 2005.

[13] Willis, N. J., "Bistatic Radar," in Radar Handbook, Third Edition, M. I. Skolnik (ed.), New York: McGraw-Hill, 2008

[14] Rittenbach, O. E., and W. Fishbein, "Semi-Active Correlation Radar Employing Satellite-Borne Illumination," *IRE Transactions on Military Electronics*, April–July 1960, pp. 268–269.

[15] Lyon, E., "Missile Attack Warning," Chapter 4 in *Advances in Bistatic Radar*, N. J. Willis and H. D. Griffiths, (eds.), Raleigh NC: SciTech Publishing, 2007.

[16] Memorandum, Chief of Naval Research to Chief of Naval Operations, Subject: CW transmit site at Spruce Creek, FL, April 29, 1966.

[17] Nicholas, R. G., "The Present and Future Capabilities of OTH Radar," *Studies in Intelligence*, Vol. 13, No. 1, Spring 1969, pp. 53–61, Central Intelligence Agency (declassified).

[18] Schoenenberger, J. G., and J. R. Forrest, "Principles of Independent Receivers for Use with Co-Operative Radar Transmitters," *The Radio and Electronic Engineer*, Vol. 52, No. 2, February 1982, pp. 93–101.

[19] Griffiths, H. D., and Long, N.R.W., "Television-Based Bistatic Radar," *IEE Proc.* Pt. F, Vol. 133, No. 7, December 1986, pp. 649–657.

[20] Ogrodnik, R. F., "Bistatic Laptop Radar: An Affordable, Silent Radar Alternative," *IEEE Radar Conference*, Ann Arbor, MI, May 13–16, 1996, pp. 369–373.

[21] Howland, P. E., "Target Tracking Using Television-Based Bistatic Radar," *IEE Proc. Radar, Sonar and Navigation*, Vol. 146, No. 3, June 1999, pp. 166–174.

[22] Howland, P. E., D. Maksimiuk, and G. Reitsma, "FM Radio Based Bistatic Radar," *IEE Proc. Radar, Sonar and Navigation*, Vol. 152, No. 3, June 2005, pp. 107–115.

[23] Sahr, J. D., and F. D. Lind, "The Manastash Ridge Radar: A Passive Bistatic Radar for Upper Atmospheric Radio Science," *Radio Science*, Vol. 32, No. 6, 1997, pp. 2345–2358.

[24] Sahr, J. D., "Passive Radar Observation of Ionospheric Turbulence," Chapter 10 in *Advances in Bistatic Radar*, N. J. Willis and H. D. Griffiths, (eds.), Raleigh, NC: SciTech Publishing, 2007.

[25] Baniak, J., et al., "Silent Sentry Passive Surveillance," *Aviation Week and Space Technology*, June 7, 1999.

[25] Griffiths, H. D., et al., "Radar Spectrum Engineering and Management: Technical and Regulatory Approaches," *IEEE Proceedings*, Vol. 103, No. 1, January 2015, pp. 85–102.

[27] https://www.asdreports.com/news.asp?pr_id=1701. Accessed September 8, 2016.

CHAPTER

2

Contents

Principles of Passive Radar

2.1 Introduction

In this chapter, the essential properties of passive radar that lead to the establishment of target detection and location performance are described. Passive radar uses illuminators of opportunity that illuminate a target or targets and capture the scattered radiation using a receiver placed remotely from the transmitter. The essential components of a passive radar system are illustrated in Figure 2.1.

In Figure 2.1, the illuminator happens to be an existing radar system and the passive receivers make up a network of bistatic adjuncts. Passive radar can use almost any type of radio frequency (RF) illuminator from a WiFi router through to space-based Global Positioning System (GPS) transmitters as well as everything and anything in between. Indeed, the research literature describes a plethora of systems composed of pretty well all-possible illuminators. As a consequence, passive radar

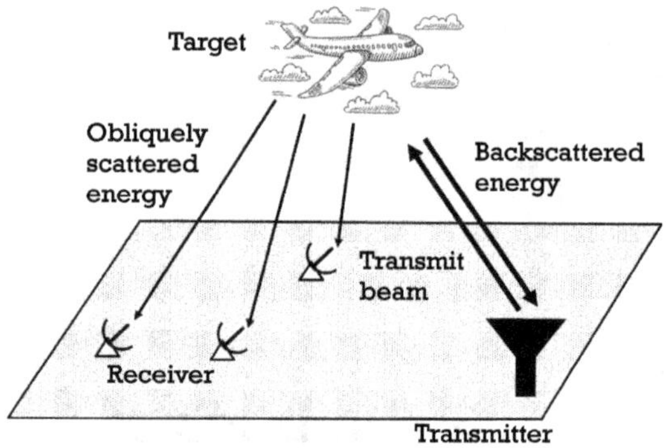

Figure 2.1 An example of a passive radar geometry.

is similar to conventional monostatic radar in that it covers a wide span of frequencies of operation, system designs, and types of application. In practice, however, there is a more restricted range of illuminators that are finding most favor. Chief among these are high-power very high frequency (VHF) radio broadcast and ultrahigh frequency (UHF) digital television (DTV) transmissions, both of which facilitate long-range aircraft detection. For this reason, VHF and UHF are mostly used as exemplars throughout this chapter when explaining the essential concepts underlying passive radar operation. However, many other types of illuminator are exploited and are described in Chapter 3. In this chapter, fundamental influences on system design and performance are introduced. Indeed, one of the most fundamental characteristics of passive radar is the wide range of transmitter-receiver geometries that are used and a description of this forms the basis of the next series of sections.

2.2 Bistatic and Multistatic Geometry

Passive radar may use one or more illuminators of opportunity and one or more receivers. Further, for reasons that will become apparent, the radar receivers are inevitably located remotely some distance away from each of the illuminating sources. Thus, for any one illuminator and any one receiver, a bistatic pair is formed and the combination of bistatic pairs comprises the overall passive radar system or network. The separation distance between an illuminator and a receiver is termed a baseline of the system. A network of illuminators and/or receivers will therefore have multiple receiver baselines. Figure 2.2 shows an example of the locations

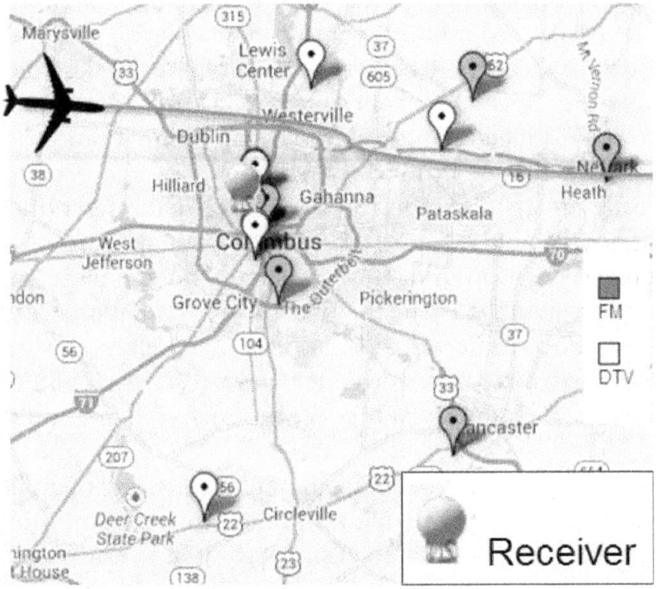

Figure 2.2 This shows the locations of the VHF FM transmitters (gray) and the UHF DTV transmitters (white) in the Columbus area together with the location of The Ohio State University passive radar receiver and a typical aircraft trajectory.

of VHF radio broadcast transmitters and UHF DTV broadcast transmitters that serve the city of Columbus, Ohio, in the United States. Drawing a line from any of the VHF or UHF transmitters to the receiver forms a baseline of the passive radar network. Therefore, the baselines spread out radially in a variety of directions and have a variety of lengths for each transmitter. In fact, the layout of transmitters and receiver positions shown in Figure 2.2 is typical of passive radar geometries where many illuminators of opportunity are available for exploitation, even by a single receiver. In other words, even if only a single receiver is used, the resulting passive radar system is, in general, multistatic.

The transmitter and receiver locations in a multistatic network are referred to as *nodes*. A single receiver keeps the passive system relatively simple and lowers the hardware and software costs. If only a single illuminator is exploited, the passive system reverts to the bistatic case and this allows basic bistatic radar theory to be applied. Further, where there are multiple illuminators, the passive system can be treated as an interconnected set of bistatic radars. This still allows simple bistatic radar theory to be used to compute the performance of the individual bistatic pairs. These individual results can then be combined to describe the overall system performance. As a consequence, in this chapter, passive radar is usually often distilled to an exemplar bistatic configuration so that fundamental aspects that determine system operation and performance can be more easily understood. The approach and bistatic descriptions loosely follow that adopted by Willis [1 (chapters 4 through 7)]. Willis provided a more detailed description of the parameters for the more general case of bistatic radar and his excellent text includes additional detail. Here, however, we concentrate on those aspects that relate most strongly to determining passive radar performance and we start with an example of bistatic geometry for a passive radar [2] as shown schematically in Figure 2.3.

The basic passive bistatic radar geometry, consisting of an illuminator of opportunity and a receiver allows the definition of a number of key terms. The transmitter and receiver are separated by the baseline L. The range from the transmitter to the target is the target range, R_T and the range from the target to the receiver is the receiver range, R_R.

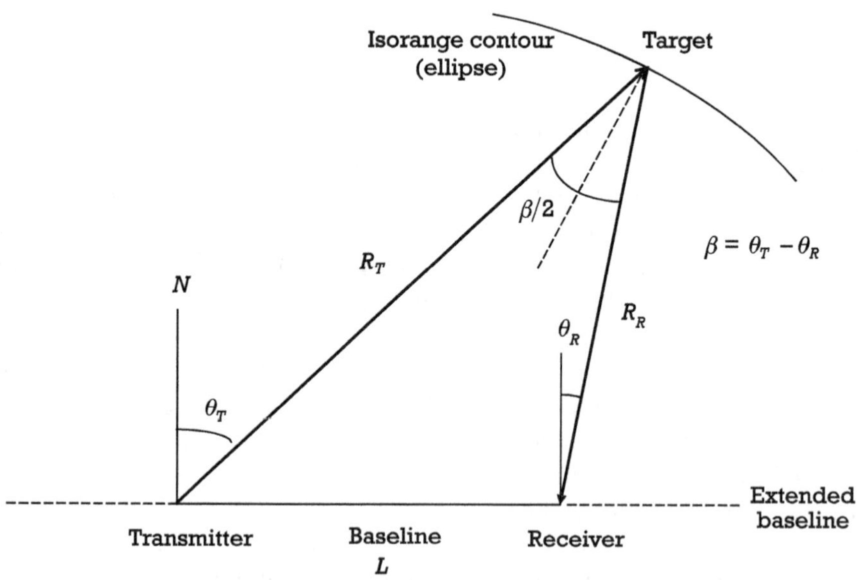

Figure 2.3 The bistatic radar geometry. The target velocity is v, making an angle δ with the bisector of the bistatic angle β.

The angle subtended at the target by the transmitter and receiver is the bistatic angle β. The bistatic bisector is the dotted line that bisects the bistatic angle. A transmitter-pointing angle, θ_T, and a receiver-pointing angle, θ_R, can be defined relative to some declared reference direction (such as north or the vertical direction as in Figure 2.3). By simple geometry this leads directly to $\beta = \theta_T - \theta_R$. For a given target velocity vector, v, the angle made between the direction of v and the bistatic bisector is denoted by δ. These parameters fully define the geometry for a single bistatic component of a multistatic passive radar system where the transmitters and receivers are in fixed locations. Note that it is possible to generalize further and allow for moving transmitters and receivers. However, this is an additional complication that is set aside to concentrate on the fundamentals and that are consistent with the majority of passive radar systems that exist today.

2.2.1 Coverage

The intersection of the illuminating transmission and the receiving antenna defines the coverage of a passive radar system. This is a fundamental aspect of system performance and a prime consideration for any prospective application. Here, the coverage provided by passive radar is reviewed in terms of the types of illuminator of opportunity most typically utilized. Passive radar illuminators of opportunity tend to direct their power towards their user community. For example, TV and radio broadcast illuminators have their users located anywhere on the ground and hence the illumination pattern is close to omnidirectional in azimuth and as much as possible of the radiation is directed downwards and outwards. However, the relatively low frequencies used in the VHF and UHF bands means that, inevitably, significant radiation leaks in an upwards direction thus making them suitable for applications such as air traffic management and air defense. In fact, many illuminators of opportunity tend to be omnidirectional in azimuth, transmitting their signal equally in all directions. However, there are other illuminators, such as scanning radars, that are highly directional but distribute their illumination over all azimuth angles as a function of scan angle.

2.2.2 Direct Signal Suppression

Because the omnidirectional illuminators, such as VHF radio and UHF TV stations transmit in all azimuth directions, there is unavoidably some of the transmitted signal that will arrive directly at the receiver without having been reflected by a target. This is called the direct signal and it is transmitted along the baseline as defined by the passive radar geometry (Figure 2.3). The direct signal is only subject to a one way propagation loss and hence is attenuated by $1/L^2$, where L is the baseline distance between the transmitter and the receiver. Thus the received signal can be very strong. This can be both an advantage and a disadvantage in the design and operation of passive radar. It is an advantage because, by receiving the direct signal through a separate antenna and receiving system (the direct channel), it can be used as a timing reference to compute the bistatic delay or range, that is, the distance from the illuminator to the target and then to the receiver. This use of the direct signal as a timing reference is more broadly referred to as "coherent on receive." It allows the phase of the target signal in the surveillance channel to be referred to a starting phase derived from the direct channel. This means that phase changes due solely to target and clutter motion can be retrieved and used in a manner similar to that of conventional, coherent, pulse Doppler monostatic radar. Indeed, as will be seen shortly, passive radar is often highly reliant on phase to detect and resolve targets by virtue of Doppler resolution rather than range resolution.

The presence of the direct signal becomes a disadvantage when it leaks into the antenna being used to detect targets (i.e., in the surveillance channel) as its strength is usually greater than that of the weak echoes from targets. The direct signal is often so strong that, even after steps have been taken to reduce it in the surveillance channel, it still represents a source of interference that competes with weak target echoes, and hence can limit the maximum detection range. The reception of the direct signal in the surveillance channel is termed *direct signal interference* and has to be reduced to a level that, in the ideal case, is below that of receiver noise. If this can be achieved, it avoids any reduction in the maximum detection range of a target of given radar cross-section. Methods for reducing direct signal interference in the surveillance channels are considered in more detail in Chapter 4.

2.3 Bistatic Range and Doppler

The measurement of range and Doppler frequency and the ability to resolve in range and Doppler are fundamental properties of any radar system and are central to the determination of detection and tracking performance. The same is true for passive radar, and hence, in this section, range, Doppler frequency, and resolution in range and Doppler are examined. Examples are given to illustrate the relative roles these parameters play in passive radar and how differences from the more conventional monostatic radar case come about.

2.3.1 Range Measurement

Referring back to Figure 2.3, the range or delay time between the directly received signal and the echo from a target is given by (2.1) and is termed the bistatic range or equivalently bistatic delay

$$R_T + R_R - L \tag{2.1}$$

$R_T + R_R$ is termed the *range sum*. The bistatic range (or delay) is a fundamental measurement made by any passive radar. Indeed, there are essentially three parameters that the bistatic receiver can measure:

1. The difference in range between the direct signal and the transmitter-target-receiver path (the bistatic range);
2. The Doppler shift f_D of the received echo;
3. If a directional surveillance channel antenna is used, the angle of arrival θ_R of the received echo.

Contours of the constant bistatic range ($R_T + R_R$) define an ellipse, with the transmitter and receiver located at the two focal points. This is in contrast to monostatic radar where the coincidence of the transmitter and receiver means that lines of constant range are circles. If the baseline, L is known, the range sum ($R_T + R_R$), can be extracted from the measureable quantity ($R_T + R_R - L$). If θ_R is measured, the range of the target from the receiver may be found via simple geometry and is given by (2.2) in terms of the known and measured quantities:

$$R_R = \frac{\left(R_T + R_R\right)^2 - L^2}{2\left(R_T + R_R + L\sin\theta_R\right)} \tag{2.2}$$

For the vast majority of the passive radar systems reported in the research literature, omnidirectional illumination sources (in azimuth) are exploited and the directionality of the receive antenna is usually quite coarse (many tens of degrees). This is primarily due to the low transmission frequencies requiring large physical structures to provide high directionality. Consequently, this means that the range from the target to the passive radar receiver can be measured accurately, but that the direction to the target may only be known to a coarse level, perhaps only of the order of a single azimuth quadrant.

The accuracy of the range measurement is determined by a combination of the effectiveness of the coherent-on-receive technique, the bandwidth of the transmitted signal, the signal-to-noise ratio at the receiver and any influence of the propagation environment.

2.3.2 Range Resolution

Following the approach adopted in [1, chapter 7], the range resolution of a conventional monostatic pulsed radar is given by $\Delta R = c\tau/2$, where c is the velocity of propagation and is the compressed pulse length. This can be rewritten as $\Delta R = c/2B$, where B is the bandwidth of the transmitted signal. Thus, in monostatic radar systems, targets can be separated in range via a series of concentric circles that have a separation of ΔR.

In passive radar, the bistatic geometry leads to a set of concentric ellipses and this difference to monostatic radar has to be additionally taken into account in evaluating passive radar range resolution. If two targets are positioned on a line that is an extension of the bistatic baseline and they are on consecutive bistatic ellipses separated by the range resolution, their resolvability is the same as for monostatic (assuming the signal bandwidths are the same). However, if they are at any other, arbitrary position away from the baseline but are colinear with respect to the bistatic bisector (see Figure 2.4), then the expression for range resolution approximates to:

$$\Delta r = \frac{c}{2B\left(\frac{\cos\beta}{2}\right)} \tag{2.3}$$

However, this is still a special case and places an unnecessary restriction on the definition of bistatic range resolution. The more general situation is one where the two targets are not colinear with the bistatic bisector and a more general but still approximate expression for passive radar resolution is given by:

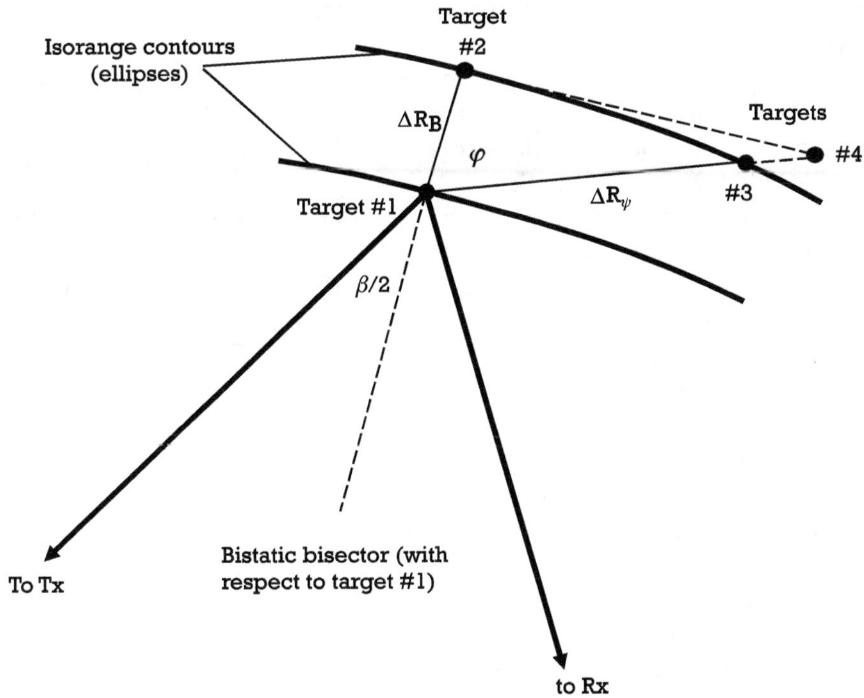

Figure 2.4 Bistatic range resolution geometry.

$$\Delta r = \frac{c}{\left[2B\left(\frac{\cos\beta}{2}\right)\right]\cos\varphi} \tag{2.4}$$

where φ is the angle between the bistatic bisector and a line joining the targets as shown in Figure 2.4. Note that (2.4) still reverts to the monostatic equivalent when the two targets lie on a projection of the bistatic baseline.

In general, however, it can be concluded from (2.4) that the range resolution in a bistatic radar geometry is coarser than in the equivalent monostatic case (assuming the signal bandwidths are the same). Further, it can be seen from (2.4) that the bistatic resolution varies as a function of the relative target position with respect to the transmitter and receiver positions. When the bistatic bisector angle is very small, such as might be the case for a long-range operation or closely spaced transmitter-receiver pairs, the ellipse begins to approximate a circle and the bistatic system becomes closer to that of a monostatic system. This can be an advantage in terms of being able to adopt many of the concepts and understanding of the operation of monostatic radar, but the advantages and differences that result from a bistatic radar geometry will become eroded. When the bistatic bisector angle is significant, such as at short ranges or for very long baseline lengths, the curvature of the ellipses is very pronounced and the bistatic geometry begins to have a dominating effect on detection performance which deviates more and more from that of monostatic radar.

Finally, when the bistatic bisector angle approaches 90° the range resolution, ΔR approaches infinity; in other words, all resolution is lost. This situation is known as *forward scatter* and is usually treated quite separately. However, for a passive radar network, a target entering the forward scatter regime for one transmitter-receiver pair will still be in a bistatic geometry for another transmitter-receiver pair. Overall, there is a continuum of resolutions from the monostatic case through to the forward scatter geometry. For practical purposes, a rule of thumb can be adopted where the resolution limits are set to an upper bound of twice the monostatic equivalent. Beyond this, the forward scatter geometry begins to dominate and system detection may revert to another transmitter-receiver pair only. More advanced passive radar concepts may attempt to process data on a continuous basis rather than consider the bistatic and forward scatter geometries as two separate regimes of operation.

In passive radar, bandwidths and therefore range resolutions can also vary considerably. For example, older analog VHF signal bandwidths are typically only a few hundreds of kilohertz and offer very coarse range resolution of the order of several kilometers. More modern digital signals, for example, high-definition television (HDTV) signals, have much higher bandwidths, in the region of 6 to 8 MHz giving range resolutions of the order of 20m. The HDTV signals are typically high power (up to 1 MW in the United States, for example) and allow much greater levels of target resolution based upon range and velocity than their VHF counterparts. This becomes especially important if systems, as is often the case, use little in the way of receive antenna resolution. In the case of VHF, there has to be much greater reliance on Doppler alone.

2.3.3 Doppler Measurement

In the general case when the transmitter, the target, and the receiver are all moving, the Doppler shift on the echo is obtained from the rate of change of the transmitter-target-receiver path. If the transmitter and receiver are stationary and referring back to Figure 2.3, the Doppler shift of the received echo is given by

$$f_D = \frac{2v}{\lambda}\cos\delta\cos(\beta/2) \tag{2.5}$$

It can be seen that, if the target is crossing the bistatic baseline in a forward scatter geometry, $\beta = 180°$ and $f_D = 0$, no matter what the direction or the magnitude of the target velocity. Physically, this can be understood by recognizing that at the point when a target crosses the baseline, the transmitter-to-target range is changing in an equal and opposite way to the target-to-receiver range. Conversely, lines of maximum Doppler are orthogonal to the directions of lines of constant range (which, as described in Section 2.2.3, are ellipses). This means that in passive bistatic radar, the lines of maximum Doppler will take the form of hyperbolae. This is in contrast to a monostatic geometry where lines of constant range are circles and lines of maximum Doppler are therefore radii. However, despite being a little more complex, lines of maximum Doppler and constant range are described by well-known mathematical formulations and are fully deterministic.

A good test for equations describing passive radar is that they revert to the monostatic case if the bistatic angle is set to zero. For example, (2.5) reverts to the monostatic case when $\beta = 0$. Note also that when $\delta = \pm\beta/2$, the target is moving towards either the transmitter or the receiver. Further, when $\delta = \pm 0°$ or $180°$, the target is moving towards or away from the baseline at the angle of the bistatic bisector.

2.3.4 Doppler Resolution

Doppler resolution is determined by the integration time, T, as is the case for monostatic radar. The relationship between integration time and Doppler resolution is an inverse one, that is, the longer the integration time, the finer the Doppler resolution. Passive radar is often operated in a staring mode where the target is illuminated on a continuous basis and hence echoes are continuously received. This facilitates long integration times to be chosen thus allowing very fine Doppler resolutions. Again, following the approach of Willis [1], this can be expressed as shown in

$$\left|f_{Tgt1} - f_{Tgt2}\right| = \frac{1}{T} \tag{2.6}$$

where $f_{Tgt1} = 2(V/\lambda)\cos\delta_1 \cos(\beta/2)$ and $f_{Tgt2} = 2(V/\lambda)\cos\delta_1 \cos(\beta/2)$.

If the two targets are assumed to be colocated, as shown in Figure 2.5, such that they share the same bistatic bisector, then we have:

$$\Delta V = \left(V_1 \cos\delta_1 - V_2 \cos\delta_2\right) \tag{2.7}$$

$$\Delta V = \frac{\lambda}{\left[2T\cos(\beta/2)\right]} \tag{2.8}$$

where ΔV is the difference between the two target velocity vectors projected onto the bistatic bisector that will allow the two targets to be distinguished based upon their different relative velocities.

As passive radar is a staring illumination system, integration times can be selected to be very long, often of the order of 1 second, hence giving a very fine Doppler resolution of the order of 1 Hz. This is exploited in narrowband VHF passive radar systems to help to remove clutter and allow multiple targets to be separately observed. It also helps overcome the relative lack of signal bandwidth and resulting coarse range resolution. The wider bandwidths and higher-range resolutions of HDTV signals may require integration times to be restricted to avoid the target traversing

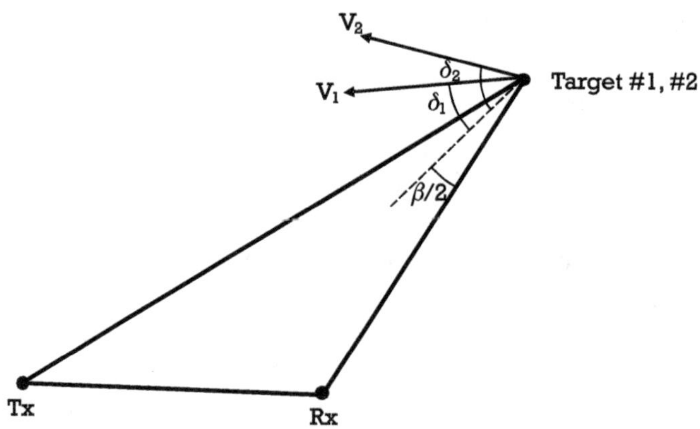

Figure 2.5 Bistatic Doppler resolution geometry.

multiple range cells. However, this can be potentially mitigated by more sophisticated processing approaches that are designed to cope with this range cell migration.

Both VHF and UHF transmissions can also be exploited to observe micro-Doppler reflections from moving parts of an air vehicle such as propeller blades and jet engine turbine blades. Figure 2.6 shows an example of a Cirrus SR 22 turboprop aircraft signature measured using a single bistatic pair comprising part of a UHF DTV passive radar geometry. The figure shows a range-Doppler map with the aircraft traveling at a velocity of approximately 65 m/s and at a bistatic range of 4 km. The micro-Doppler scattering caused by the propeller blades manifests itself as sidebands either side of the aircraft echo and the spacing of the sidebands provides information about the rotation rate of the propellers. Note also that the clutter has a spatial extent of around 1 km and a Doppler span of around ±20 m/s. The Doppler span is caused primarily by a combination of ground moving targets and zero Doppler ground clutter spillover. The integration time in this example was 0.1 second, giving a fine Doppler resolution of 10 Hz. This enables the scattering from the moving parts of

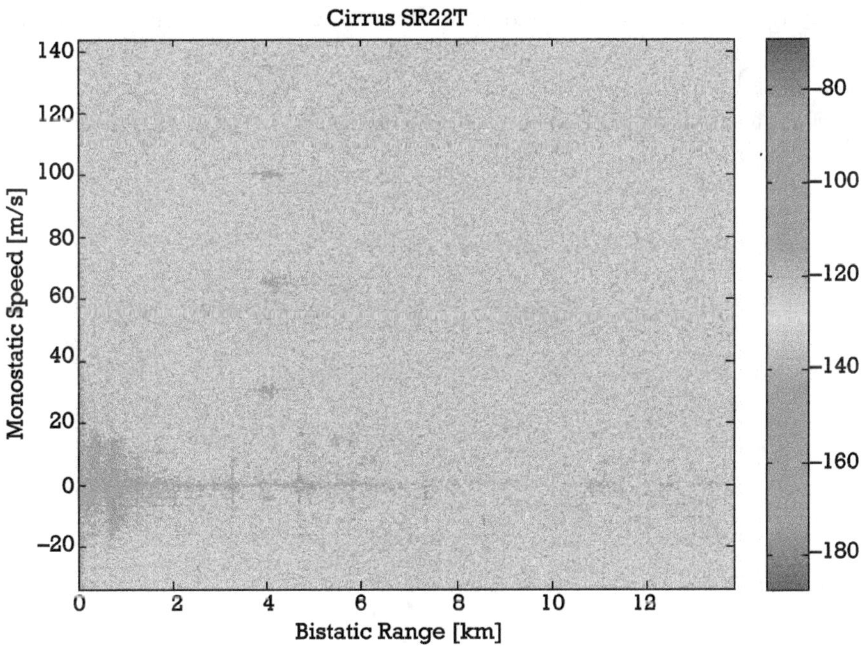

Figure 2.6 A range-Doppler map of passive radar echoes from a Cirrus SR22 aircraft showing Doppler sidebands due to the effect of the rotating propeller.

the aircraft to have significant integration gain thus allowing them to be more easily observable in Figure 2.6.

2.4 Multistatic Passive Radar Range and Doppler

A multistatic passive radar configuration has multiple transmitters and receivers placed at a number of distributed locations. A passive radar network combines multiple bistatic measurements of targets that are viewed at different aspects and the subsequent extraction of information about targets should be improved. Such a network, broken down into a series of connected and cooperating bistatic components, allows all of the explanations of range and Doppler described in the earlier part of this chapter to be applied so that systems can be designed to deliver best performance for a chosen application.

Figure 2.7 illustrates a very simple multistatic system with two transmitters and a single receiver. Because each bistatic system provides a different aspect view of the target, the multistatic system will provide simultaneous multiple bistatic views (two in this example) in a range, Doppler, angle resolution cell. As we saw in the previous section, the range from the receiver to a target is determined by a pair of transmit and receive nodes and is obtained by measuring the time delay Δt between the transmitted and the received signal $|R_t + R_R| - L = c\Delta t$.

To maximize the sensitivity of a multistatic passive radar system, it is necessary to coherently combine the signals received from each of the

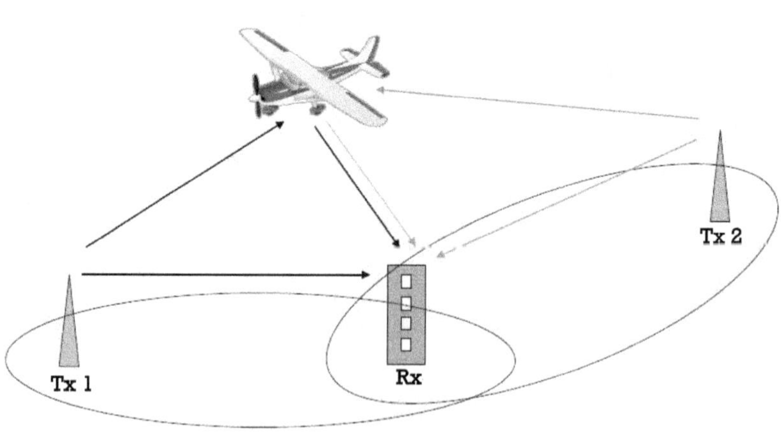

Figure 2.7 Multistatic radar system with two transmitters and one receiver.

bistatic components such that they follow the change of the target position with time allowing the evolving phase change in each bistatic pair to be computed. Assuming that the received signals in each channel can be coherently combined, the total baseband signal can be expressed by:

$$s_r(t) = \sum_{n=1}^{N} A_n \exp\{-j\Phi_n(t)\} = \sum_{n=1}^{N} A_n \exp\{-j\frac{2\pi f[R_{T,n}(t) + R_{R,n}(t)]}{c}\} \quad (2.9)$$

where N is the total number of channels, An is the amplitude and $\Phi_n(t)$ is the phase function of the signal in the nth channel, $R_{T,n}(t) = |R_{T,n}(t)|$ is the distance from the nth transmitter to the target, and $R_{R,n}(t) = |R_{R,n}(t)|$ is the distance from the target to the nth receiver.

Coherent combination in passive radar networks is complex and often not an option as the transmissions are generally not designed for synchronized operation. However, it may be possible to accomplish this in single frequency transmitter networks. When using widely spaced transmissions at differing frequencies, the individual range and Doppler measurements can still be fused to provide additional information for detection and location of targets. This is the subject of ongoing research.

As mentioned earlier, some targets such as jet engine aircraft, turboprop aircraft, and helicopters can impart a micro-Doppler modulation due to the rotating or moving components. Taking the time derivative of the combined baseband signals leads to an expression for the multistatic, micro-Doppler

$$f_{mD_{Multi}}(t,P) = \sum_{n=1}^{N} A_n f_{mD_{Bi}}(t,P) \quad (2.10)$$

In a multistatic configuration, each node has its own individual geometry between the transmitter, receiver and each target. Consequently, the multistatic topology determines the multistatic micro-Doppler features. The combination of the number of nodes and the angular separation between these nodes determine the micro-Doppler signature of the target. However, it should also be recalled that, generally, the multistatic Doppler is less than that of the monostatic counterpart and this can reduced the detail observable in a passive radar system and additionally, makes the micro-Doppler a function of geometry. Further, the long wavelengths typical of many passive radar systems operating in the VHF and

UHF bands, means that the micro-Doppler generated by targets can be much less that at higher transmission frequencies.

2.5 Multistatic Target Location

Multistatic target location is important aspect of performance and is one that has to be treated with care in passive radar. This is especially so, as noted in an earlier section, because many transmitters are omnidirectional. This means that location has to be done on receive only and may require high gain and therefore, at low frequencies, large physical structures. An alternative is to use a low-gain receive antenna with an auxiliary antenna and to employ techniques such as direction or time difference of arrival. Very broad transmit and receive beam pairs means that targets will not be resolved in angle and gives rise to ambiguities, especially when large numbers of targets are present.

In practice, the reliability and accuracy of a passive radar system can be very significantly improved by exploiting multiple transmitters and, optionally, multiple receivers as well. The location of the target measured by the bistatic radar is ambiguous and for omnidirectional transmit and receive antennas, is constrained to an ellipse with focal points at the position of the transmitter and the receiver. However, the second bistatic pair will also, in the same way, provide a measurement of range. A single intersection of the two ellipses representing the ambiguous range in each bistatic pair will remove the ambiguities in range and allow the target to be correctly located. This topic is covered in more detail in Chapter 6.

The problem of locating a target and associating measurements is eased if three or more transmitters are used as is often the case for passive radar, especially in more densely populated urban areas where multiple transmitters are often available. If this is not the case, then multiple receivers can be employed. If, for example, there are three transmitters illuminating a target, the three ellipses will usually only ever intersect in one location. The target's location can thus be easily found, even without a measurement of bearing. The problem is then a matter of finding the locations at which three or more ellipses (one from each transmitter) intersect. This is somewhat eased with HDTV signal where the wider bandwidths and commensurately high-range resolutions better separates closely spaced targets in range and ambiguities are consequently reduced. However, overall, there are no ready solutions to this problem and individual systems tend to adopt bespoke approaches.

2.6 The Bistatic Radar Range Equation

The radar range equation is a useful tool in radar design as it links the specification of the main operating parameters with detection performance in a concise fashion. However, do note that the radar range equation should be used as a guide only and is merely a start for more detailed designs.

Before the bistatic and multistatic forms of the radar equation are introduced, some key components of a passive radar system are briefly introduced (and will be considered in more detail in Chapter 3). These are all associated with the form of the signal emitted by an illuminator of opportunity and comprise: (1) coverage (i.e., the area or volume that the emitter of opportunity illuminates), (2) transmitted power, and (3) design of the emitted waveform.

Together, coverage, transmitted power, and waveform design provide information basic to the detection and tracking of targets. They are all very much dependent on particular transmitters and as transmitters can vary so much, so does the resulting performance. For example, VHF and UHF transmitters are popular for use in passive radar as they are generally high powered, typically from 10 kW to 1 MW. In addition, as already noted, they are omnidirectional in azimuth and in elevation are biased power towards the Earth's surface. This is good for low-level performance, detecting those targets that might otherwise fly under the radar. It is less attractive for detecting very high-flying aircraft if there is insufficient radiation that is transmitted in an upward direction allowing aircraft at cruising altitudes to be detected at long ranges. While the coverage and power levels show comparability between VHF and UHF transmitters, the design of their waveforms is very different. They are both continuous wave (CW) but frequencies and modulation structures are very dissimilar. VHF waveforms generally sit between 80 and 110 MHz, whereas UHF are from 200 to 800 MHz. VHF waveforms are usually analog, although there is a hybrid analog-digital waveform that is broadcast in the United States and Canada. They also occupy quite a narrow band from 200 to 400 kHz of spectrum. UHF DTV are, as the name implies, digital and have bandwidths between 6 and 7 MHz, giving a best range resolution of the order of 25m. The suitability of the waveforms for radar applications is also part determined by the modulation structure and this can be investigated further using the bistatic form of the ambiguity function. This is considered in more detail in Chapter 3.

There is a range of frequencies and bandwidths since different countries set different standards for the broadcast signals. Power levels differ depending on location and the need to reach customers. These types of illuminator are also subject to continual change. Relatively recently, in parts of Europe, China, and Australia, radio content is increasingly broadcast digitally in the form of digital audio broadcasts (DAB). DAB is transmitted in the band between 174 and 240 MHz and there are a number of variants. Power levels tend to be below that of VHF FM radio as digital encoding offers improved sensitivity after reception. The digital encoding uses orthogonal frequency demultiplexing and spans a bandwidth of approximately 1.5 MHz giving a best range resolution around 100m.

A very different type of illuminator is provided by an existing radar system. For example, in developed countries there are networks of such radars providing services for air traffic management and air defence that provide wide area coverage with radiation deliberately designed to illuminate the sky. Transmitted powers are high and the waveforms are clearly suited to radar use. However, these radars are generally mechanically scanned and this can place more exacting requirements on the remote receiver used to form the passive radar. If the receiver uses an omnidirectional antenna, these requirements are relaxed but at the expense of system sensitivity (0-dB receive antenna gain) and a higher likelihood of ambiguity (no azimuth resolution).

In general, there are advantages and disadvantages to all potential illuminators and therefore individual case studies are required to determine if available illuminators are matched to a required application. In addition, use of the electromagnetic spectrum is constantly changing and this means that passive radar design should avoid relying on any particular signal. In fact, there is no reason to choose a single transmitter type and operation at, for example, VHF simultaneously with UHF is perfectly possible and provides additional design freedoms to enhance performance.

Lastly, it must also be remembered that transmitters of opportunity are designed and installed by a third party. As a consequence, this part of the passive radar cannot usually be altered and provision of transmission cannot be guaranteed. This represents a significant aspect of design restriction that has to be carefully factored into an assessment of the suitability of passive radar to any given application. This restriction should also be tempered against the advantages of a multistatic, multifrequency transmitter networks that are often available to the passive radar designer.

Once the properties of available transmitters have been established, the radar equation can be used to relate the major design parameters to expected performance just as for any other form of radar. The starting point for an analysis of the performance of a simple single transmitter, single receiver passive radar system is the basic form of the bistatic radar equation:

$$\frac{P_R}{P_n} = \frac{P_T G_T}{4\pi R_T^2}.\sigma_b.\frac{1}{4\pi R_R^2}.\frac{G_R \lambda^2}{4\pi}.\frac{1}{kT_0BF} \tag{2.11}$$

where P_R is the received signal power, P_n is the receiver noise power, P_T is the transmit power, G_T is the transmit antenna gain, RT is the transmitter-to-target range, σ_b is the target bistatic radar cross section, RR is the target-to-receiver range, G_R is the receive antenna gain, λ is the signal wavelength, k is Boltzmann's constant, 1.38×10^{-23} W/K/Hz, T_0 is the noise reference temperature, 290K, B is the receiver effective bandwidth, and F is the receiver effective noise figure.

Equation (2.11) can be modified by including loss terms, pattern propagation factors, and processing gain, but the simple form illustrates its basic properties and dependences.

The factor $\left(\frac{1}{R_T^2 R_R^2}\right)$ in (2.11) means that the signal-to-noise ratio has a minimum value when $R_T = R_R$, and is greatest for the two cases when the target is either very close to the transmitter or very close to the receiver. Contours of constant values of $\left(\frac{1}{R_T^2 R_R^2}\right)$, and hence of signal-to-noise ratio, define geometric figures known as *Ovals of Cassini* [1,2]. An example of the Ovals of Cassini for a passive radar is shown in Figure 2.8. This assumes omnidirectional transmit and receive antennas; for directional antennas the contours are weighted by the antenna patterns and may take quite different shapes.

It should be noted that the fact that there are Ovals of Cassini describing contours of constant signal-to-noise ratio and elliptical lines describing contours of constant range means that the relationship between signal-to-noise ratio and target location is no longer one for one as in the monostatic case. Again this is a minor complication but a completely deterministic one.

The basic form of the radar range equation (2.10) is modified by the inclusion of losses, by pattern propagation factors on the transmitter-to-target and target-to-receiver paths, and by appropriate integration gain. The noise figures of receivers, F, at VHF and UHF will be of the order of a

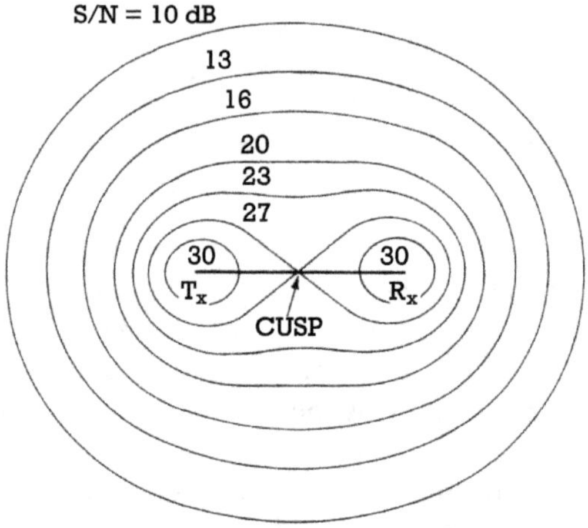

Figure 2.8 Ovals of Cassini representing contours of constant signal to noise ratio for a passive bistatic radar geometry [1].

few decibels, so the noise level will be dominated by external noise, most likely in the form of the direct signal, multipath, and other cochannel signals. Unless steps are taken to suppress these signals the sensitivity and dynamic range of the system will be severely limited. This is considered in more detail in Chapter 4. High-power VHF and UHF transmitters enable detection ranges against large aircraft to be in the region of al hundred km or more. Equation (2.11) is quite general and can be used to predict performance for all types of illuminators. However, for passive bistatic radar, there are some additional uncertainties that have to be considered. Not least of these are the magnitude and form of bistatic scattering from targets, a key component of the radar range equation.

2.7 Bistatic Target and Clutter Signatures

In general, the bistatic radar cross-section of a given target will not be the same as its monostatic radar cross-section, though for a nonstealthy target the values may be comparable. However, complex targets such as aircraft will scatter incoming radiation at all angles and hence the bistatic radar cross-section will be a strong function of target type and bistatic geometry. Early in the history of bistatic radar the bistatic equivalence theorem was put forward [3]. The equivalence theorem states that the bistatic radar cross-section of a given target at a bistatic angle β is equivalent to the monostatic radar cross-section measured at the bisector of the bistatic angle, reduced in frequency by the factor cos ($\beta/2$), provided that the target is sufficiently smooth, there is no shadowing of one part of the target by another, and retroreflectors persist as a function of angle. In practice, these conditions may not always be met, so the theorem should be used with care, especially for large bistatic angle and for complex targets. However, as a starting point for computations at the radar equation level, it should prove adequate.

Note that bistatic operation can be a counter to stealth technology. This is because stealth technology has been designed to combat monostatic radar and does this, in part, by minimising target reflections back in the direction of the incoming illuminating radiation. However, some of the energy must be scattered at other angles, potentially giving some advantage to a bistatic geometry In addition the lower frequencies in the VHF band, especially, are also used to counter stealth and hence passive radars could experience a double advantage.

To gain further insight into the difference that can occur between monostatic scattering and bistatic scattering and how geometry plays an important role, we now consider a couple of almost trivial examples using simple scatterer types. Many practical targets will include a combination of flat plate, dihedral, and trihedral scatterers that may result in a high monostatic radar cross-section compared to the bistatic radar cross-section. Here we consider a simple square flat plate target to illustrate some of the differences in scattering magnitude that can ensue.

Figure 2.9 shows the geometry with two monostatic radars illuminating a flat plate target such that the bistatic pairs of transmitters and receivers are in a specular condition where the angle of incidence equals

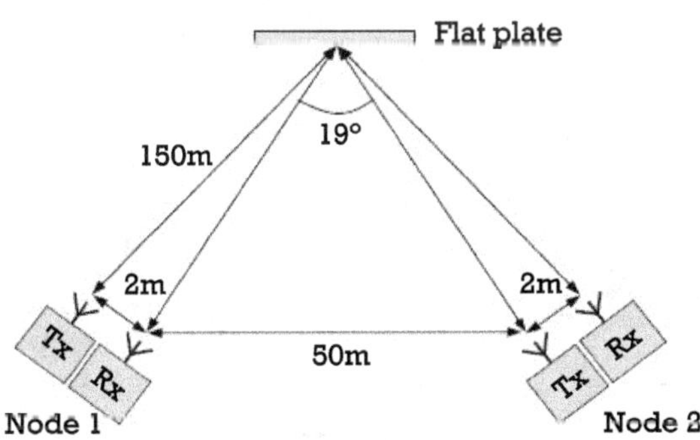

Figure 2.9 A dual monostatic bistatic measurement geometry. The flat plate would need to be rotated by 19° (in either direction) to maximize the monostatic echo response.

the angle of reflection. For a monostatic radar there is a maximum radar cross-section when the radar views the flat plate in a direction normal to the plane of the plate. This radar cross-section maximum reduces as the plate is rotated away from the orthogonal plane such that the response declines by half the maximum receive power (i.e., 3 dB) at an angle approximately given by λ/d, where λ is the wavelength of the illuminating radiation and d is the length of one of the sides of the square plate. As the plate rotates to larger angles away from the orthogonal position, the magnitude of the scattering back in the direction of the radar continues to fall rapidly and then rise and falls in decreasing amounts in a manner consistent with that described by a sinc function (again as for a uniformly illuminated antenna). With bistatic illumination, the received signal is always small except at the specular condition as shown in Figure 2.9 and also produces a very different response function compared to the monostatic case. Figure 2.10 shows how much larger the bistatic scattering is than for the monostatic geometry, which illuminates the flat plate at an angle of 19° away from the orthogonal plane. However, the likelihood of the specular condition being met is quite low in practice and more generally the bistatic radar cross-section of targets is lower than that of the monostatic counterpart.

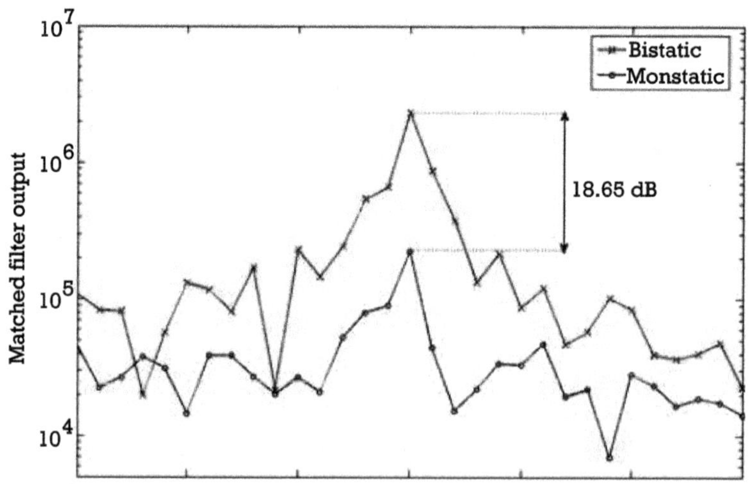

Figure 2.10 The radar cross-section of a flat plate viewed at an angle of 19° to normal incidence by a monostatic radar and for a specular geometry for the bistatic radar.

Now we consider replacing the flat plat by a cylindrical target. The response is the same for both the bistatic and monostatic radars regardless of the illumination geometry. Together, these two, almost trivial, examples using very simple targets illustrates the range of different behaviors that are possible and will be a feature of scattering from more complex real targets. Measurements of the radar cross-section of real targets are very few and far between with some exceptions such as the radar cross-section of ships as a function of bistatic angle. These measurements are consistent with the expected complexity of scattering. Figure 2.7 shows one of the rare examples of published data comparing monostatic target signatures with bistatic target signatures, in this case for ship targets [4]. The geometry is one where a ground based monostatic radar and a bistatic based adjunct receiver was used and the targets were ships at sea.

Figure 2.11 shows the results for four ships. It is clear from the figure that the bistatic radar cross-section is generally smaller than the monostatic radar cross-section and that this difference increases as a function of increasing bistatic angle. Thus, when computing performance using the radar range equation, it would be prudent to use a range of radar cross-section values to indicate the range of possible performance as a function of geometry changes as a target moves through the surveillance area.

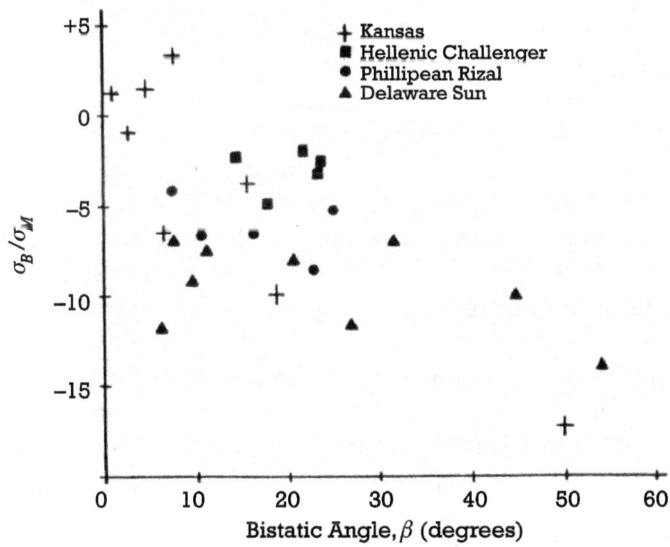

Figure 2.11 A plot of the ratio of the median bistatic radar cross-section to the monostatic radar cross-section as a function of bistatic angle, β for four small ship targets [4].

Although these measurements were made using a dedicated illuminator operating at X-band, the trends, overall, are indicative of what might be expected more generally for passive radar. However, there have been very few reports of bistatic radar cross section measurements using passive radar and this is an aspect of passive radar that is currently very poorly understood. There has also been little published in the literature relating to statistical target models. However, the well-known Swerling models can be employed for computation of detection and false alarm probabilities.

Glint is a target (and clutter) scattering phenomenon that may also be substantially reduced in bistatic geometries. The reasons for this come from the basic understanding of bistatic scattering as described above. Glint is a change in the apparent direction of a wavefront caused by constructive and destructive interference from scatterers comprising a target. Consequently, it tends to be dominated by larger echoes. However, as has been seen, the bistatic cross-sections generally are smaller than their monostatic counterparts and hence it might be expected that glint is also reduced. However, again, the experimental evidence to support such a hypothesis is scant and there are no reported results explicitly for passive radar. Indeed, there is only one example for bistatic radar that is referred to in [1].

There are three mechanisms that have been put forward that might enhance the bistatic radar cross-section of a target but under particular conditions. These are resonant scatter, specular scatter, and forward scatter. The first two of these are also effective in monostatic geometries.

Resonant scatter occurs when physical dimensions of the target (such as the length of an engine or the distance between nose and wing root of an aircraft) correspond to multiples of half of the radar wavelength and can also be understood with reference to the classical frequency dependence of scattering from a conducting sphere. In general, such effects will be dependent both on frequency and on target aspect. For example, at VHF wavelengths are in the region of a few meters and it is easy to suppose that large-scale features such as those mentioned above will resonate. At UHF where wavelength might vary from around 150 to 30 cm, it will be correspondingly smaller-scale objects that will resonate with, in general, correspondingly smaller radar cross-section values. At still higher frequencies, the resonant effect becomes less of a dominating factor in determining target backscatter.

As described above, *specular scattering* will occur if the target possesses flat features that happen to be oriented such as to give a specular reflection. However, such scattering depends on the specular condition being met and will consequently be somewhat rare and random.

It was seen earlier that *forward scatter* occurs when a target crosses the baseline between the transmitter and the receiver. According to Babinet's principle, the signal diffracted around a target of a given silhouette area will be equal and opposite to that diffracted through an equivalent target-shaped hole in an infinite screen perpendicular to the path between transmitter and receiver [5]. The signal diffracted through an aperture of a given shape and area can be readily calculated, allowing the forward scatter radar cross-section to be determined directly. For a target of silhouette area A and linear dimension d, the forward scatter radar cross-section is approximately $\sigma_{FS} = \frac{4\pi A^2}{\lambda^2}$ (m^2) and the angular width of the forward scatter is approximately $\theta_B \approx \frac{\lambda}{d}$ (radians). These are plotted in Figure 2.6 for a target for which $A = 10$ m^2 and $d = 10$m (a medium-sized aircraft), and it can be seen that the forward scatter radar cross-section can be substantially greater than the equivalent monostatic radar cross-section (which for this target might be of the order of 10 m^2). However, it can be appreciated that while this geometry may give good detection performance, the target location capability will be poor, since the range and Doppler resolution will both be poor when the target is near the baseline. This point is developed further in the next chapter through the development and analysis of the bistatic ambiguity function.

From Figure 2.12, it can be seen that the optimum frequency that provides a large radar cross-section over a reasonably broad a range of viewing angles is a little above 300 MHz, right in the middle of the UHF DTV band. However, Figure 2.12 is constructed from a very simple repre sentation of forward scatter and the radar cross-section enhancement can be observed without too much difficulty over a broad range of frequencies. As passive radar typically consists of a number of bistatic pairs, the likelihood of targets crossing baselines is seemingly quite high; however, this has to be considered carefully for any given locations of transmitter and receiver, together with the trajectory of the target. Consider an example with a ground-based transmitter operating at a frequency of 350 MHz and receiver separated by a baseline of 50 km. An aircraft at a cruising altitude of 10 km crosses the baseline halfway between the transmitter and the receiver. When the aircraft crosses the baseline, it will make

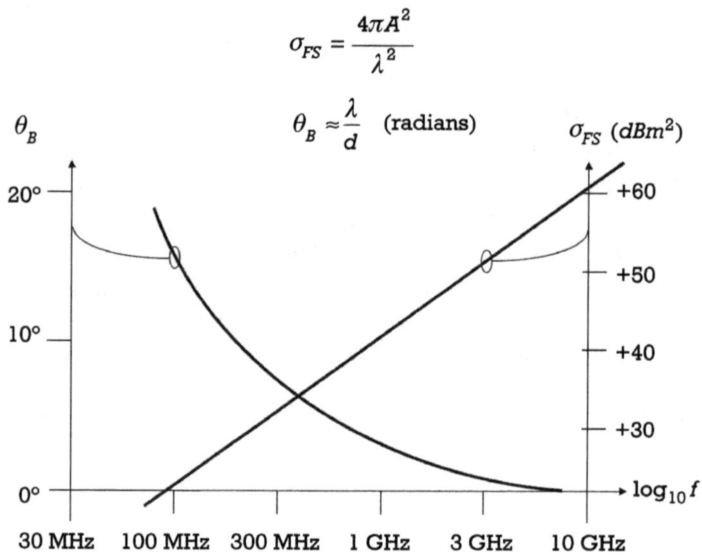

Figure 2.12 Forward scatter radar cross-section σ_{FS} and angular width of scatter θB for an idealised medium aircraft target with $A = 10$ m^2 and $d = 10$m.

an angle of over 20° as seen by the receiver. This places it outside of the observable region as indicated by Figure 2.12. Thus, forward scatter using ground-based transmitters and receivers might be better suited to the detection of low-flying aircraft, which may be of high interest in defense or security applications. This is especially so, given that there can be a significant radar cross-section enhancement that defeats conventional stealth technology.

Finally, the role of clutter should be mentioned. Defining clutter loosely as unwanted reflections it is desirable to both know about the properties of clutter and also to be able to remove clutter. Currently, there is no comprehensive understanding of clutter as manifest in passive radar systems and suitable models have yet to be developed. Again, the bistatic geometry can be invoked but the few reported results for bistatic operation that exist [6] do not cover many of the frequencies most often used in passive radar (i.e., VHF and UHF). However, as can be observed in Figure 2.6, clutter, especially at short ranges can be significant having an echo strength that inhibits effective target detection. Just as with any form of radar, clutter has to be understood and accounted for in both performance modeling and the design of signal processing.

2.8 Summary

This chapter has introduced the essentials of passive radar and combining them via the radar range equation to relate parameter specification to detection performance. The importance of controlling direct signal breakthrough has been highlighted as has the relatively poor understanding of bistatic scattering from targets (and clutter). Equally, we have seen that the performance of a bistatic radar system is very heavily influenced by the choice of illuminator, which dictates frequency of transmission, coverage, and waveform modulation. Despite the restrictions that come from using existing illuminators, passive radar systems have been successfully designed, constructed, and demonstrated to have remarkably high levels of performance. Their relative simplicity, low cost, and spectral efficiency has made them the subject of great interest. However, much less is understood about bistatic radar systems than their monostatic counterparts and they have some particular differences in their design and operation. These differences form the subject of following chapters in which the topics introduced here are examined in greater detail.

References

[1] Willis, N. J., *Bistatic Radar*, 2nd ed., Silver Spring, MD: Technology Service Corp., 1995; corrected and republished by Raleigh, NC: SciTech Publishing, 2005.

[2] Jackson, M. C., "The Geometry of Bistatic Radar Systems," *IEE Proc.*, Vol. 133, Pt. F, No. 7, December 1986, pp. 604–612.

[3] Kell, R. E., "On the Derivation of Bistatic RCS from Monostatic Measurements," *Proc. IEEE*, Vol. 53, August 1965, pp. 983–988.

[4] Ewell, G. W., and S. P. Zehner, "Bistatic Radar Cross Section of Ship Targets," *IEEE J. Oceanic Engineering*, Vol. OE-5, No. 4, October 1980, pp. 211–215.

[5] Born, M., and E. Wolf, *Principles of Optics*, 6th ed., London, U.K.: Pergamon Press, 1980, p. 559.

[6] Weiner, M., "Clutter," Ch. 9 in *Advances in Bistatic Radar*, N. J. Willis and H. D. Griffiths, (eds.), Raleigh, NC: SciTech Publishing, 2007.

CHAPTER

3

Contents

Properties of Illuminators

The choice of illuminating source is a key factor in determining the performance of a passive radar system. There are several parameters to be taken into account when assessing the utility of such illuminators. The first of these is the power density (in W/m^2) of the illuminator at the target. The significance of this is apparent in considering the detection performance, and is discussed in Chapters 2 and 5. The second is the nature of the waveform, and the third is the coverage, and these factors are discussed in this chapter.

3.1 Ambiguity Functions

Radar signals, indeed, signals of all kinds, can be expressed as a function of time or of frequency, the two representations being related via the Fourier transform. Repetitive features in one domain will result in distinct features in the other. The performance of a radar waveform is classically determined by its ambiguity function. This was originated by the British mathematician Philip Woodward in the 1950s [1]. Formally it is defined as the square magnitude of the output from a filter matched to the transmitted signal st(t), and represents the response of the radar to a point target as a function of delay T_R and Doppler shift f_D

$$\left|\psi\left(T_R, f_D\right)\right|^2 = \left|\int_{-\infty}^{\infty} s_t(t) s_t^*\left(t+T_R\right) \exp\left[j 2\pi f_D t\right] dt\right|^2 \tag{3.1}$$

The width of the peak of the ambiguity function shows the resolution of the radar in range and in Doppler, and the ambiguity function also shows the sidelobe structure and any ambiguities that may result from periodic features of the waveform. In a conventional pulsed radar, this will show the ambiguities in range and velocity associated with the pulse repetition frequency (PRF), spaced at $c/(2PRF)$ and $(\lambda PRF)/2$, respectively.

3.1.1 The Ambiguity Function in Bistatic Radar

In a bistatic radar, the ambiguity function depends not only on the waveform, but also on the bistatic geometry, in other words, the position of the target with respect to the transmitter and receiver. This can be understood by realizing that for a target on the baseline between transmitter and receiver, there is no resolution in range, because the echo arrives at the receiver at the same time as the direct signal, irrespective of the target location. Also, there is no resolution in Doppler, because for a target crossing the baseline the transmitter-to-target range changes in an equal and opposite way to the target-to-receiver range, so the Doppler shift is zero irrespective of the target velocity. Equally, it can be appreciated that in a bistatic radar, the simple linear relationships that exist in a monostatic radar between target range and delay, and between target velocity and Doppler shift, are more complicated.

These factors mean that in a bistatic radar the ambiguity function depends on a greater number of variables, and should instead be written [2] as:

$$\left|\psi\left(R_{RH},R_{Ra},V_H,V_a,\theta_R,L\right)\right|^2 = \left| \begin{array}{l} \int_{-\infty}^{\infty} s_t\left(t-\tau_a\left(R_{Ra},\theta_R,L\right)\right)s_t^*\left(t+\tau_R\left(R_{RH},\theta_R,L\right)\right) \\ \times \exp\left[\begin{array}{l} j2\pi f_{DH}\left(R_{RH},V_H,\theta_R,L\right) \\ -2\pi f_{Da}\left(R_{Ra},V_a,\theta_R,L\right)t \end{array}\right]dt \end{array} \right|^2 \tag{3.2}$$

where R_{RH} and R_{Ra} are the hypothesized and actual ranges from the receiver to the target, V_H and V_a are the hypothesized and actual target radial velocities with respect to the receiver, f_{DH} and f_{Da} are the hypothesized and actual Doppler frequencies, θ_R is the north-referenced direction of arrival of the target echo, and L is the bistatic baseline.

A practical passive radar system will need to take this effect into account, but because it is deterministic, if a target is being detected and tracked by a number of bistatic transmitter-receiver pairs, it can be realized when the target is close to the baseline on one such pair and the information from that pair either discarded or suitably weighted in forming the track. These ideas are developed in Chapter 6.

It is readily possible to measure and plot the ambiguity functions of potential passive radar sources by receiving and digitizing off-air signals, and this has been done by many researchers [3–5]. Typically, such measurements can use a spectrum analyzer in the zero scan mode as a versatile receiver or a software-defined radio module. In all cases, these show the monostatic ambiguity function.

Figure 3.1 presents some typical results of this kind. The width of the peak of the ambiguity function in range is related to the instantaneous waveform bandwidth B, so $\Delta R = c/2B$ and the width in Doppler is $\Delta_{fD} = 1/T$, where T is the integration time, which is limited in practice by the target coherence time as explained in Chapter 5.

Figure 3.1(a) shows the ambiguity function of a very high frequency (VHF) frequency modulation (FM) radio station broadcasting a signal with speech modulation (BBC Radio 4). The peak and the sidelobe structure are well defined, although the peak is relatively broad, as a consequence of the low spectral content of the modulation. Figure 3.1(b) shows the equivalent result for an FM radio station with fast-tempo jazz music modulation (Jazz FM). The peak and the sidelobe structure are

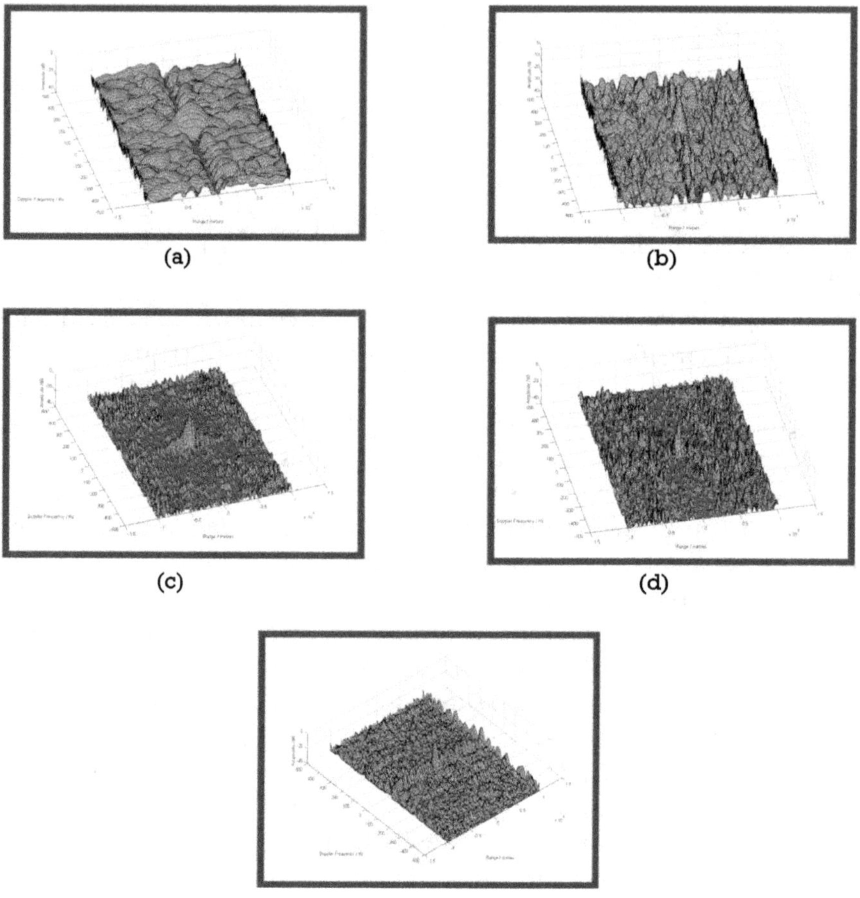

Figure 3.1 Measured ambiguity functions from (a) VHF FM BBC Radio 4 (speech), (b) VHF FM Jazz FM (fast tempo jazz music), (c) Digital Audio Broadcast (DAB) at 222.4 MHz, and (d) Digital Video Broadcasting (DVB-T) at 505 MHz, and GSM900 at 944.6 MHz.

correspondingly sharper because of the higher spectral content of the modulation. In both cases, the floor of the ambiguity function is down by a factor of $(B\tau)^{1/2}$, rather than by $(B\tau)$, which would be expected for coherent waveforms.

Figures 3.1(c–e) show typical ambiguity functions for digital transmissions (DAB, DVB-T, and GSM, respectively). These functions are more favorable for PBR purposes than signals with analog modulation such as

Figures 3.1(a, b), since the peak of the ambiguity function is narrower and the sidelobes are lower. Also, their form is time-invariant and does not depend on the program or information content.

Table 3.1 summarizes the properties of various types of illuminator source. In each case the power density has been calculated at a typical value of transmitter-to-target range, and with the assumption of free space line-of-sight propagation. In the same way as with a monostatic radar, the range resolution is dictated by the signal bandwidth (though

Table 3.1
Summary of Typical Parameters of PBR Illuminators of Opportunity

Transmission	Frequency	Modulation, Bandwidth	P_tG_t	Power density (note 1) $\Phi = \frac{P_tG_t}{4\pi R_T^2}$
HF broadcast	10–30 MHz	DSB AM, 9 kHz	50 MW	–67 to –53 dBW/m^2 at R_T = 1,000 km
VHF FM	88–108 MHz	FM, 200 kHz	250 kW	–57 dBW/m^2 at R_T = 100 km
Analog TV	~550 MHz	Vestigial sideband AM, 5.5 MHz	1 MW	–51 dBW/m^2 at R_T = 100 km
DAB	~220 MHz	Digital, OFDM, 220 kHz	10 kW	–71 dBW/m^2 at R_T = 100 km
DVB-T	~750 MHz	Digital, 6 MHz	8 kW	–72 dBW/m^2 at R_T = 100 km
Cell phone base station (GSM)	900 MHz, 1.8 GHz	GMSK, FDMA/TDMA/FDD, 200 kHz	100 W	–71 dBW/m^2 at R_T = 10 km
Cell phone base station (3G)	2 GHz	CDMA, 5 MHz	100W	–71 dBW/m^2 at R_T = 10 km
WiFi 802.11	2.4 GHz	DSSS/OFDM, 5 MHz	100 mW	–41 dBW/m^2 (note 2) at R_T = 10m
WiMAX 802.16	2.4 GHz	QAM, 1.25–20 MHz	20W	–88 dBW/m^2 at R_T = 10 km
GNSS	L-band	CDMA, FDMA, 1–10 MHz	200W	–134 dBW/m^2 at Earth's surface
DBS TV	Ku-band, 11–12 GHz	Analog and digital	300 kW	–107 dBW/m^2 at Earth's surface
Satellite SAR	9.6 GHz	Chirp pulse, 400 MHz (max.)	28 MW	–54 dBW/m^2 SAR (note 3) at Earth's surface

Source: [6]. Note 1: assuming free space line-of-sight propagation; Note 2: would be subject to additional attenuation due to propagation through walls; and Note 3: parameters from SARs carried by the COSMO-SkyMed series of satellites [7].

also by the bistatic geometry), and the typical bandwidth of each kind of signal is given in the third column.

3.1.2 Bandwidth Extension with FM Radio Signals

Although the range resolution corresponding to the bandwidth of a single FM radio channel is rather coarse ($B = 50$ kHz gives $c/2B = 3$ km, and the actual value may be even coarser depending on the program content), it is possible to exploit several channels broadcast by a single transmitter to achieve better resolution. The concept was first proposed by Tasdelen and Köymen [8], who showed that using seven adjacent FM channels improved the range resolution by a factor of about 3, although at the expense of introducing some range ambiguities.

Bongioanni [9], Olsen [10–12], and Zaimbashi [13] took the idea further and verified its operation experimentally. The basic processing scheme used by Olsen is shown in Figure 3.2. Two receivers are used, one for the direct signal and one for the target echoes. In each receiver, the whole band is digitized, and each FM is signal filtered and then upconverted to yield adjacent channels at the same spacing Δf. Next, the cross-correlation operation is performed. The results in [12] demonstrate that range resolutions as high as 375m may be obtained, although the results

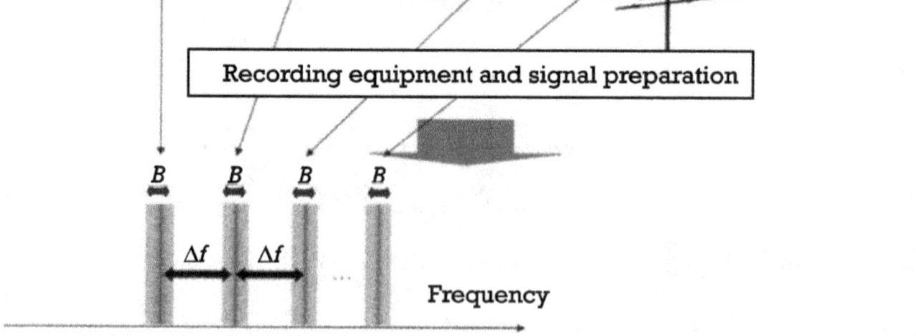

Figure 3.2 Processing scheme for bandwidth extension with FM radio signals [11].

still depend on the bandwidths (and hence the program content) of the individual stations.

Because the channels have different carrier frequencies, the Doppler shifts for a given target velocity will be different for each channel, which in practice limits the coherent integration time. Olsen's algorithm compensated for the different Doppler shifts emerging from different carrier frequencies from the same target, as well as independent phase terms, which allowed stable coherent integration, and hence gain and Doppler resolution. The work was further extended to cope with range migration [14] and was demonstrated using data from three DVB-T channels (not adjacent and not regularly spaced) integrated coherently for 4 seconds.

3.2 Digital Versus Analog

The results in Figure 3.1 show some important distinctions between the performance in radar terms of digital modulation formats compared to analog. With analog signals, the ambiguity function will in general be time-varying, and will depend on the nature of the program content. For a radar signal, this is clearly undesirable.

3.2.1 Analog Television Signals

The effect is illustrated in Figure 3.3. This shows a spectrum analyzer display of an analog television signal (on the right) and the equivalent digital television signal (on the left). At the time that this result was obtained (September 2005) in the United Kingdom, both signals were being broadcast, side by side.

The modulation format of the analog color television signal is known as Phase Alternating Line (PAL), used in the United Kingdom. Other analog modulation formats that were used in other countries, such as NTSC (in North and Central America) and SECAM (in France) were broadly similar. For PAL in the United Kingdom, the picture information is made up of successive lines, each of 64-μs duration and with a 12-μs sync pulse at the beginning of each line. The whole picture is made up of 625 such lines, in two interlaced scans. This information is amplitude-modulated onto the vision carrier such that the lower sideband is suppressed and only the upper sideband is evident (vestigial sideband). The color information of the picture is modulated onto a chrominance subcarrier, and in this case the sound information is present both in analog and digital form.

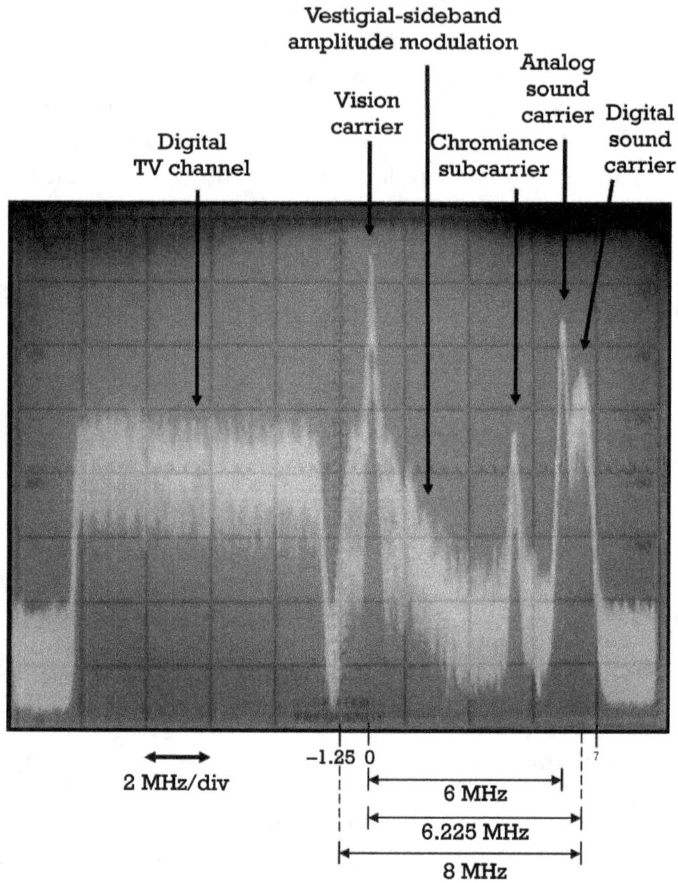

Figure 3.3 Spectrum of analog television signal (PAL) and the equivalent digital television signal (horizontal: 2 MHz/div, vertical: 10 dB/div).

When the PAL signal is used as a radar signal, there are pronounced ambiguities at 64-μs intervals, both because one line of a television picture will usually be very similar to the next, and because of the presence of the sync pulse at the beginning of each line. This results in strong range ambiguities at (in monostatic terms) 9.6-km intervals. There are also ambiguities corresponding to the frame scan rate of 25 Hz. These effects, plus the fact that the amplitude modulation of the vision carrier never actually reduces to zero, mean that the analog television signal is far from ideal as a radar signal.

The equivalent digital television signal, to the left, has a flat, noise-like spectrum, which is much more satisfactory as a radar waveform, as the ambiguity function will have a uniform sidelobe structure, which is essentially time-invariant and which is independent of the program content.

In most countries the analog TV service has been discontinued, and replaced by digital TV (Section 3.3.4). In the United States, for example, analog television was discontinued in 2009. France switched off all analog services on November 29, 2011, to be followed by Japan on March 31, 2012.

3.2.2 Mismatched Filtering

In most passive radar processing schemes the cross-correlation process represented by (3.1) is realized by employing two receiver channels: one to receive the target echoes (the signal channel), and one to receive a reference version of the transmitted signal (the reference channel). The output from the processing is therefore the cross-ambiguity function of the target echo with the reference signal:

$$\left|\psi\left(T_R, f_D\right)\right|^2 = \left|\int_{-\infty}^{\infty} s_t(t) s_r^*\left(t+T_R\right)\exp\left[j2\pi f_D t\right]dt\right|^2 \tag{3.3}$$

where $s_r(t)$ is the reference signal. The reference signal should be as clean as possible, free from multipath and with a high signal-to-noise ratio. If there is a clear line of sight from the transmitter to the passive radar receiver, this is usually straightforward, though the use of a directional antenna pointed at the transmitter will also be helpful. In some cases a clean version of the direct signal is obtained as part of the direct signal suppression processing described in Chapter 5, or in particular cases with a cooperative source the direct signal may be obtained by physical (cable) link from the transmitter, in which case it is uncontaminated by multipath.

However, it is also possible to modify the reference signal to give a mismatched filter in order to improve the cross-ambiguity function, for example, to remove sidelobes, since the features of the signal that cause the unwanted sidelobe peaks are known a priori. This approach was first proposed by Saini and Cherniakov [15], and has been applied with some success with several of the waveforms considered in the next section.

3.3 Digitally Coded Waveforms

The past decade has seen the introduction of numerous digitally coded modulation schemes for communications and broadcast applications. These include GSM, 3G and 4G cellphone signals, DAB, DVB-T and DRM broadcast signals, and WiFi (802.11) and WiMAX (802.16). Several of these are based on orthogonal frequency-division multiplexing (OFDM). Next we describe the basis of OFDM and some of these signals and their potential as radar sources.

First, it is useful to understand the nature of the communications channel, because this gives insight into why the signals are designed in the way that they are [16]. In most cases, the channel will consist of a direct path plus a number of multipath components, each with its own amplitude and delay. The direct path and the multipath components may be time-varying and Doppler-shifted. Multipath components may add destructively at the receiver, causing *fading*, and because the phase corresponding to a given delay is a function of frequency, the fading will therefore be a function of frequency.

If the delay spread associated with the multipath starts to become comparable with the bit length, *intersymbol interference* will occur and the received signal will be corrupted (Figure 3.4). This sets a limit on the data rate that be transmitted through the channel.

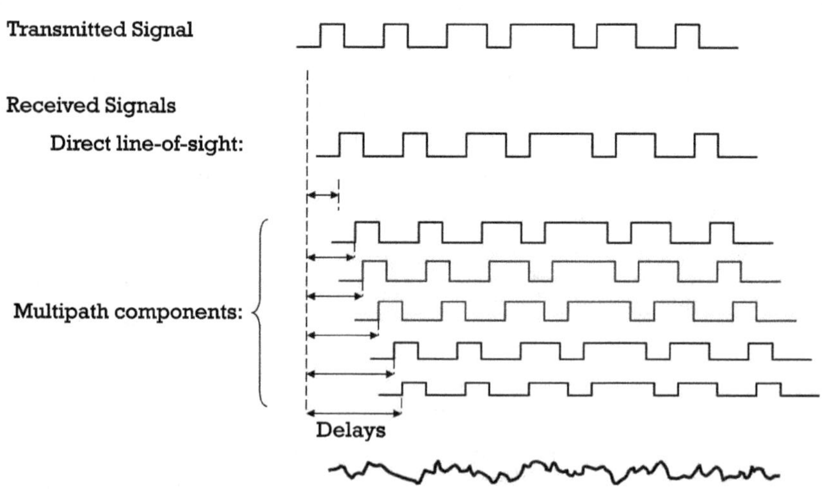

Figure 3.4 Intersymbol Interference caused by multipath.

3.3.1 OFDM

To attempt to overcome this, OFDM was devised. Here, the digital bit stream is multiplexed into a number of parallel streams, so that the bit length T_b in each individual stream is stretched by a factor equal to the number of parallel streams, N. The bit length is now much greater than the maximum delay spread of the multipath (Figure 3.5), so the multipath has relatively little effect. The parallel data strea-ms are modulated onto a set of subcarriers, spaced in frequency such that the nulls of the $(\sin x)/x$ modulated spectrum of one subcarrier are coincident with the carrier frequencies of all of the others (Figure 3.6), in other words, that they are orthogonal to each other. The subcarrier spacing in frequency required to achieve this is $1/\tau$, where τ is the bit length of the expanded bit stream.

The signal, consisting of the modulated simultaneous subcarriers, is transmitted over the channel. In the receiver each subcarrier is individually demodulated and then the original data stream is reconstituted by demultiplexing.

This form of modulation also allows the use of single-frequency networks, in which the all of the transmitters for a given station share the same frequency and are synchronized together. OFDM has been shown to be an effective means of combatting the effects of multipath, and forms the basis of many of the modulation formats used in modern communications and broadcast.

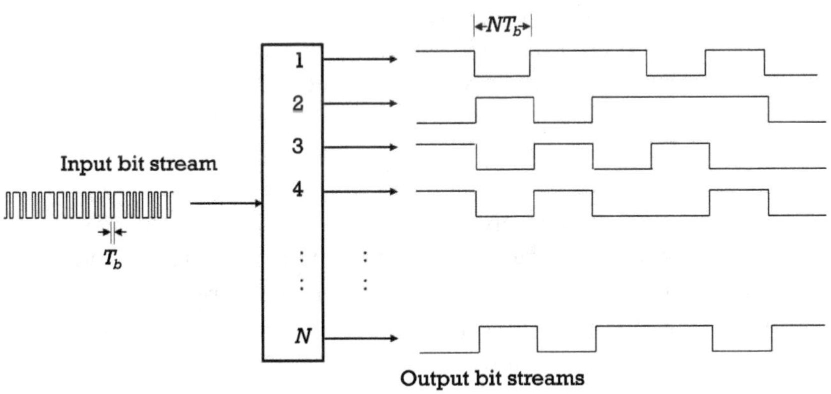

Figure 3.5 OFDM: multiplexing of the digital bit stream into multiple parallel streams. The bit length of the output bit streams is now much greater than the maximum delay spread of the multipath.

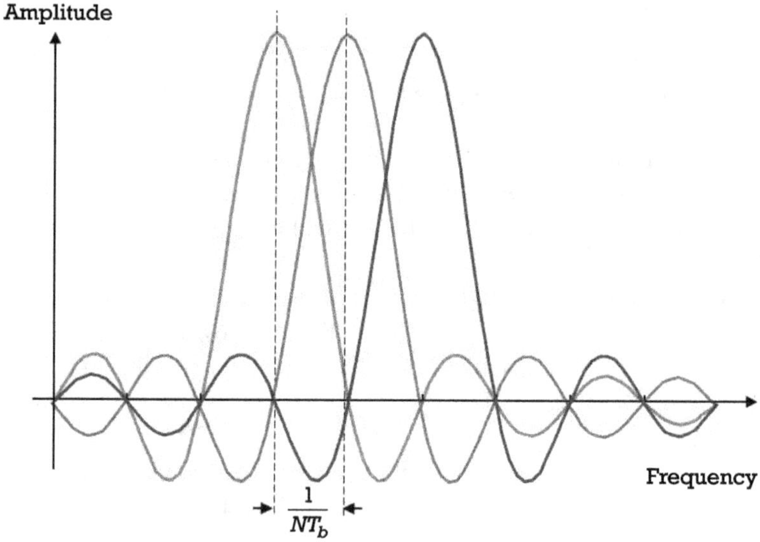

Figure 3.6 Orthogonal subcarriers in OFDM.

3.3.2 Global System for Mobile Communications

The Global System for Mobile Communications (GSM) standard was developed by the European Telecommunications Standards Institute (ETSI) for 2G cellular phone networks and is now widely adopted worldwide. It uses bands centered on 900 MHz and 1.8 GHz, and 1.9 GHz in the United States. The uplink and downlink bands are each of 25-MHz bandwidth, at 900 MHz split into 125 FDMA (frequency division multiple access) carriers spaced by 200 kHz, and at 1.8 GHz split into 375 FDMA carriers. A given base station will only use a small number of these channels. Each of these carriers is divided into 8 TDMA (time division multiple access) time slots, with each time slot of duration 577 μs. Each carrier is modulated with using Gaussian minimum-shift keying (GMSK) modulation. A single bit corresponds to 3.692 μs, giving a modulation rate of 270.833 kbps.

Figure 3.8 shows a range cut through the ambiguity function of the GSM signal of Figure 3.7. It can be seen that the periodic features, both at the time slot rate and frame rate of the time-domain signal, result in pronounced range ambiguities. However, the range resolution (~1,000m) implied by the bandwidth of the signal (~150 kHz) is too coarse for the short-range applications for which this kind of signal is suited. However, with suitable integration intervals it is possible to obtain useful discrimination in Doppler.

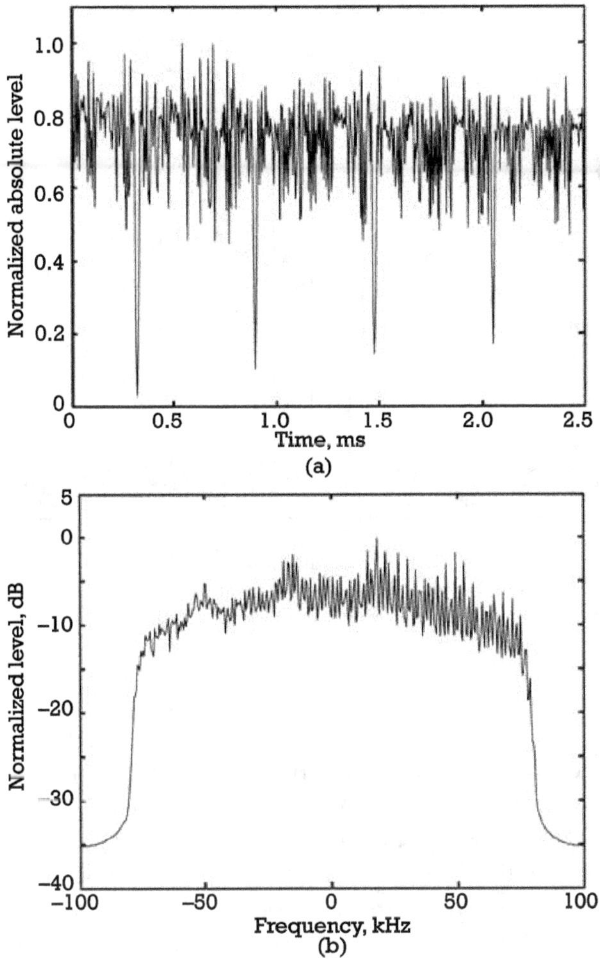

Figure 3.7 Time domain (left) and frequency domain (right) representations of GSM modulation [17].

3.3.3 Long-Term Evolution

The Long-Term Evolution (LTE) waveform is carrier modulated digital data transmitted on OFDM subcarriers. OFDM was considered in the 1990s for 3G systems, but the more mature wideband code division multiple access (WCDMA) was chosen instead. Currently, OFDM is widely used in systems such as 802.11 (WiFi), 802.16 (WiMAX), and DAB/DVB broadcasting. Its widest use is currently in the fourth generation (4G) mobile systems based on LTE. These systems started being deployed

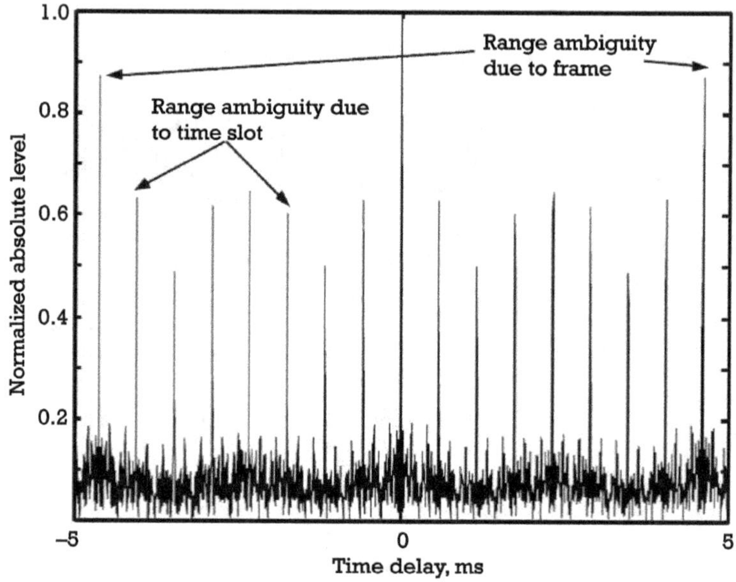

Figure 3.8 Range cut of ambiguity function of GSM signal [17].

worldwide in 2012 and are spreading widely. LTE offers data and voice services with downlink data rates up to 100 Mbps. The LTE base channel is defined according to channel bandwidths ranging from 1.4 to 20 MHz, divided across a number of OFDM subcarriers ranging from 72 to 1,320. LTE may operate across several frequency bands; 32 bands are defined by the standardization body 3GPP [18] with frequencies ranging from 729 MHz to 3.8 GHz. Operators are allocated specific spectrum bands depending on licensing and bandwidth requirements, with widths that are multiples of 5 MHz for example in the U.K. operators are allocated bands of 5, 10, 15, 20, 25, and 35 MHz [19]. LTE uses various modulation formats depending on the type of information transmitted and the quality of the wireless channel with QPSK, 16QAM, and 64QAM defined in the standard [21].

The multiple access scheme of LTE is based on OFDM and is termed OFDMA. Users are allocated bandwidth according to demand and traffic loading in a cell. The basic unit of allocation in LTE is termed the LTE Resource Block (RB) and this is based on 1 ms repeating subframe divided into two 0.5-ms slots. Each slot contains 12 subcarriers with 15-kHz fixed spacing and either 6 or 7 OFDM symbols (depending on the length of the cyclic prefix used); a single subcarrier and one OFDM symbol define what is termed Resource Element (RE), which is the smallest information unit of LTE [18, 20]. Therefore, an LTE time-domain signal is based on the aggregation of subframes into 10-ms frames.

An illustration of LTE symbols and frames and resource block is shown in Figure 3.9. An LTE resource block is defined as a 0.5-ms slot containing 7 OFDM symbols each in turn comprising 12 orthogonal subcarriers with 15-kHz spacing. In LTE downlink, resource blocks are assembled in resource grids as shown in Figure 3.10. In this representation the repetitive features are apparent.

To facilitate channel estimation and to transmit control signals, pilots, synchronization, and control channels are sent periodically, thus resulting in cyclostationary features. The cyclostationarity of OFDM and of the LTE downlink transmission has been studied and the associated signature has been analyzed in different environments (see, for example, [23, 24]). The cyclostationarity features influence the overall signal time-frequency characteristics and their efficacy for use in radar applications.

Figure 3.10 shows the spectrum of an LTE signal. The example shows a resource grid (top) and spectrum of a 1.4-MHz LTE signal having six resource blocks (i.e., 72 subcarriers) and with a simulation period of 10 ms (i.e. simulating a full radio frame comprising ten 1-ms subframes indicated by the 140 OFDM symbols on the x-axis). It can be seen that the spectrum is essentially flat and noise-like.

The corresponding ambiguity function is shown in Figure 3.11. It can be seen that there is a sharp peak at the origin and a more or less uniform sidelobe level at about –30 dB with respect to the peak, but secondary peaks are evident due to the nature of the cyclic prefix being a repeated (copied) portion of the original symbol. A similar result has been shown in [23] for the case of the extended cyclic prefix, showing a single secondary peak.

Although the ambiguity function is already quite suitable for radar purposes, current research is looking how it may be still further optimized [25, 26]. Further comments on this are made in Chapter 8.

3.3.4 Terrestrial Digital Television

Another example of the use of OFDM is in DVB-T (terrestrial digital television) signals [27, 28]. A full description of the DVB-T signal format may be found in reference [29], but a short summary is provided here. These signals use either 2k or 8k subcarriers depending on the operating mode. The OFDM symbols come from three different data streams:

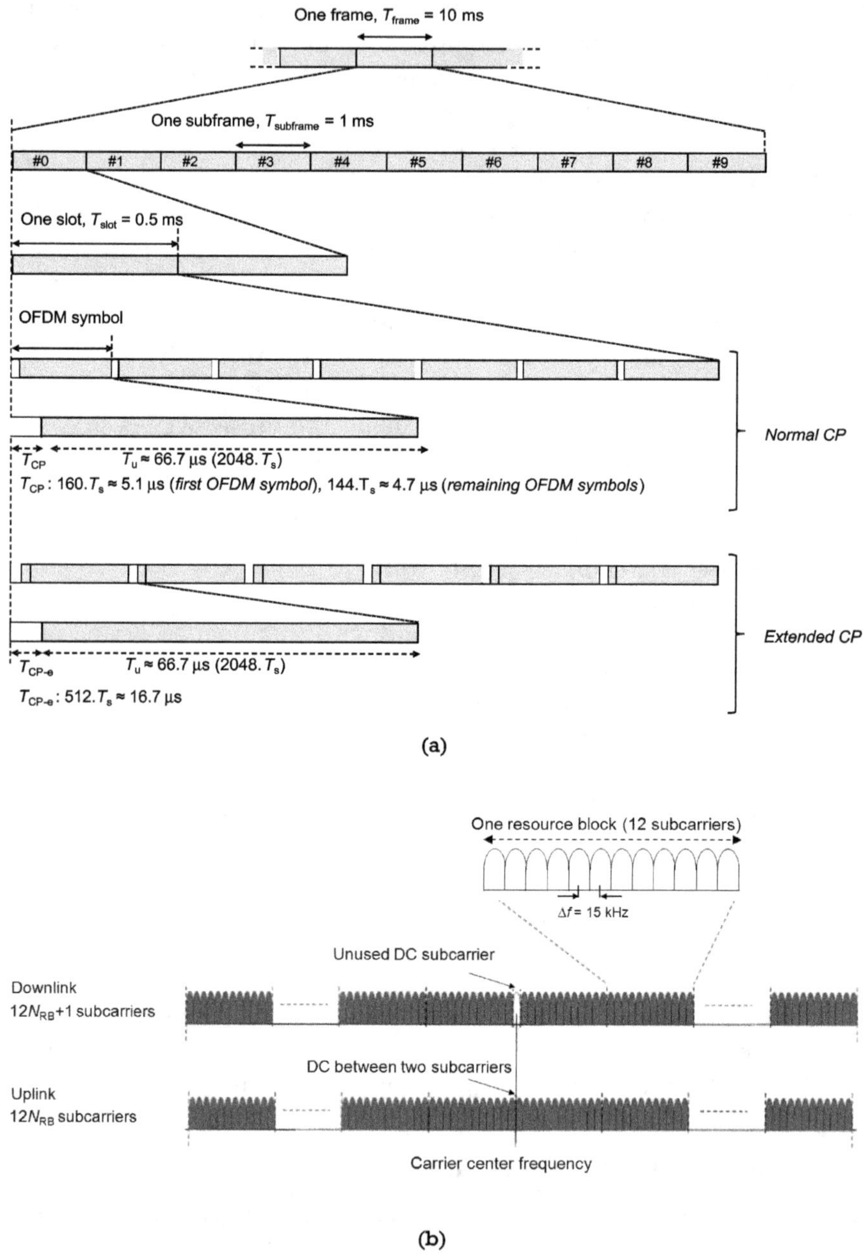

Figure 3.9 (a) LTE time domain, (b) frequency domain representations © Elsevier [20], and (c) resource block structure © 2011. 3GPP™ TSs and TRs are the property of ARIB, ATIS, CCSA, ETSI, TTA and TTC who jointly own the copyright in them. They are subject to further modifications and are therefore provided to you "as is" for information purposes only. Further use is strictly prohibited. [21].

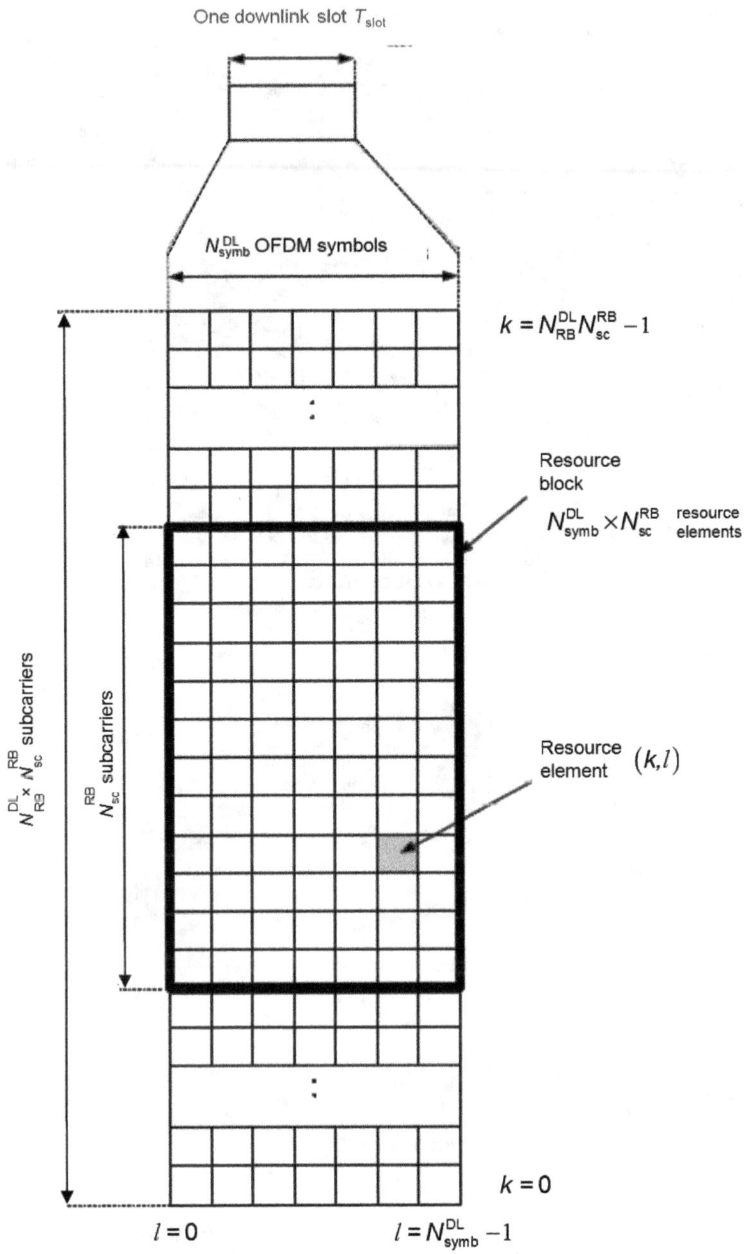

(c)

Figure 3.9 (continued)

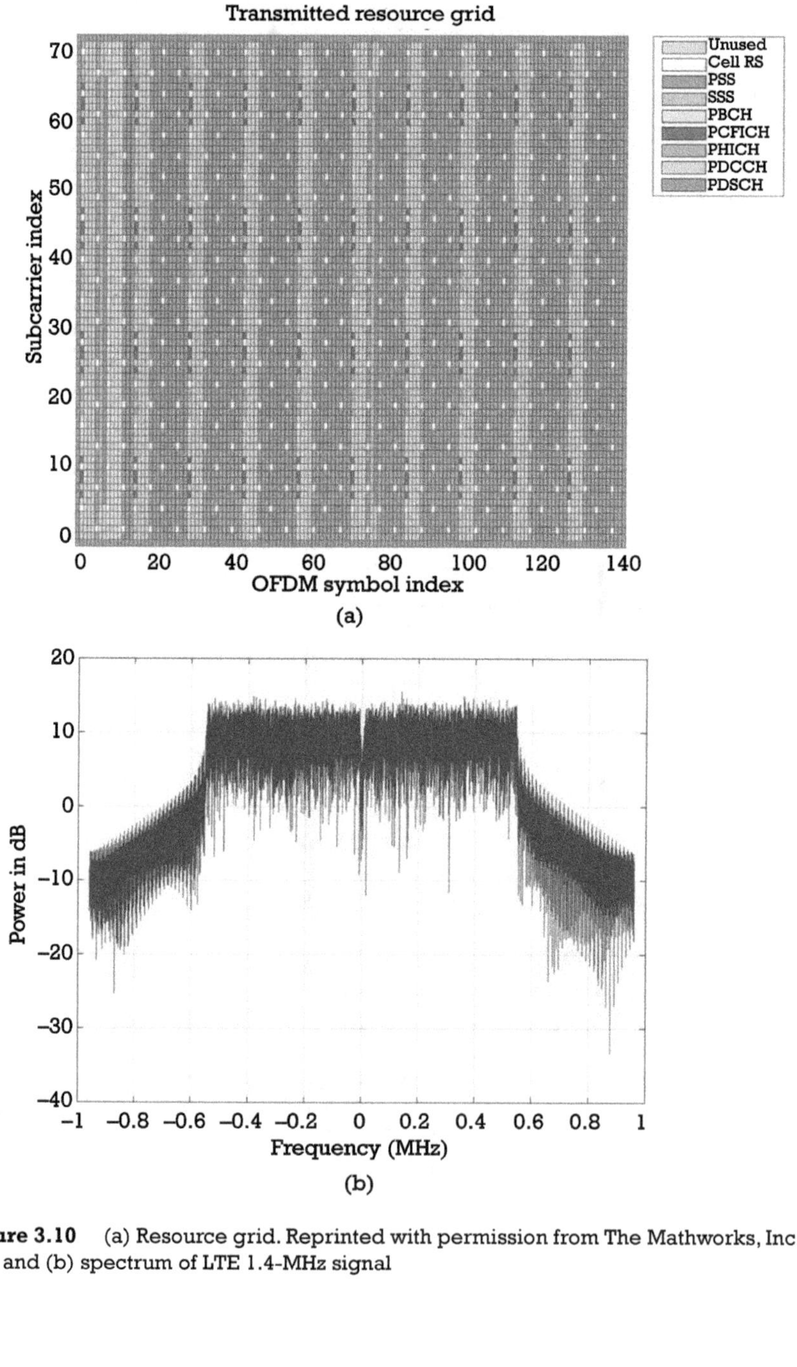

Figure 3.10 (a) Resource grid. Reprinted with permission from The Mathworks, Inc. [22] and (b) spectrum of LTE 1.4-MHz signal

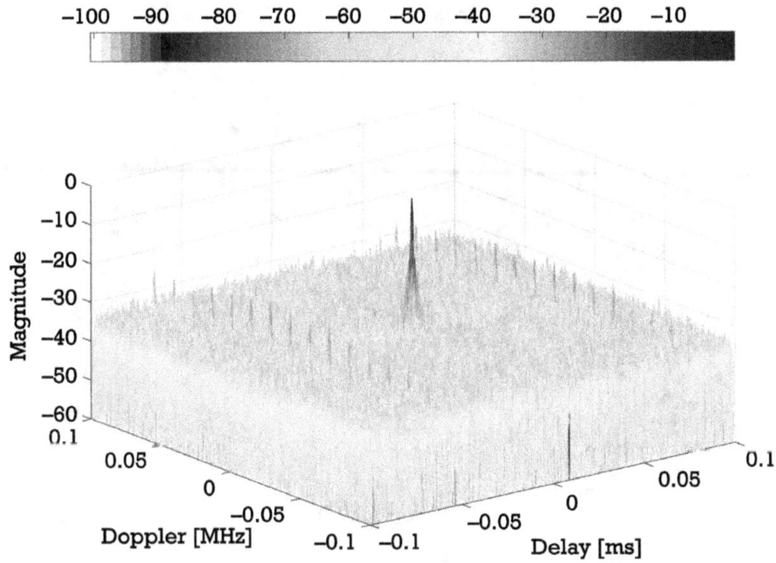

Figure 3.11 Ambiguity function of LTE signal.

1. The MPEG-2 data: The data stream is subject to bit randomization, outer coding, and inner coding and is then mapped into the signal constellation. This leads to a flat noise-like spectrum with a total bandwidth of about 7 MHz (Figure 3.12). The data carriers are modulated with QPSK, 16-QAM, or 64-QAM according to the operating mode.

2. *Transmission parameter signal (TPS):* The TPS carriers provide the transmission scheme parameters. The standard defines the locations of the carriers, which are constant.

3. Pilot definition: The pilot symbols are used by the receiver in the demodulation and decoding of the received signal. There are two types of pilots: scattered pilots, which are uniformly spaced, and continual pilots, which occupy the same carrier consistently from symbol to symbol.

The corresponding ambiguity function is similar to that of Figure 3.10, with a single narrow peak at the origin plus some smaller subsidiary sidelobes. Techniques to improve the ambiguity functions of DVB-T signals are described in [28, 30].

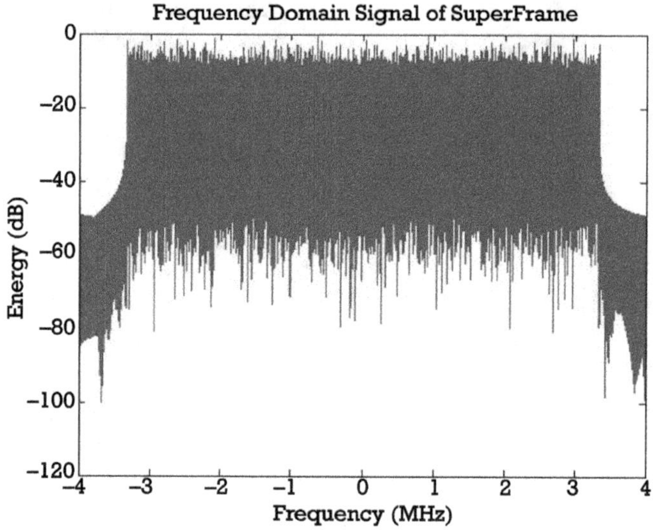

Figure 3.12 DVB-T signal spectrum [27].

3.3.5 WiFi and WiMAX

Another class of signal that has received significant attention for short-range surveillance using passive radar are the wireless transmissions for WiFi local area networks (LANs) (IEEE Std. 802.11) [31] and WiMAX metropolitan area networks (MANs) (IEEE Std. 802.16) [32, 33]. The 802.11b and 802.11g standards operate in the 2.4-GHz band, while 802.11a uses the 5-GHz band. The WiFi standard is low-power and short-range, intended primarily for indoor use and therefore potentially usable for surveillance within buildings or for short-range outdoor applications; the WiMAX standard provides broader coverage (up to several tens of kilometers) so may be useful for applications such as port or harbor surveillance.

The 802.11 WiFi modulation formats and their use as radar signals are described in detail in [34]. The resolution in range is of the order of 25m, with a peak range sidelobe level of around 18 dB. The Doppler resolution is given by the reciprocal of the integration time, with relatively high sidelobe levels, of the order of 6 dB. The EIRP will depend on the particular access point and antenna, but will be of the order of hundreds of milliwatts at most.

The use of the 802.16 WiMAX signal and its ambiguity function properties are described in [35–37]. Figure 3.13, from [37], shows the

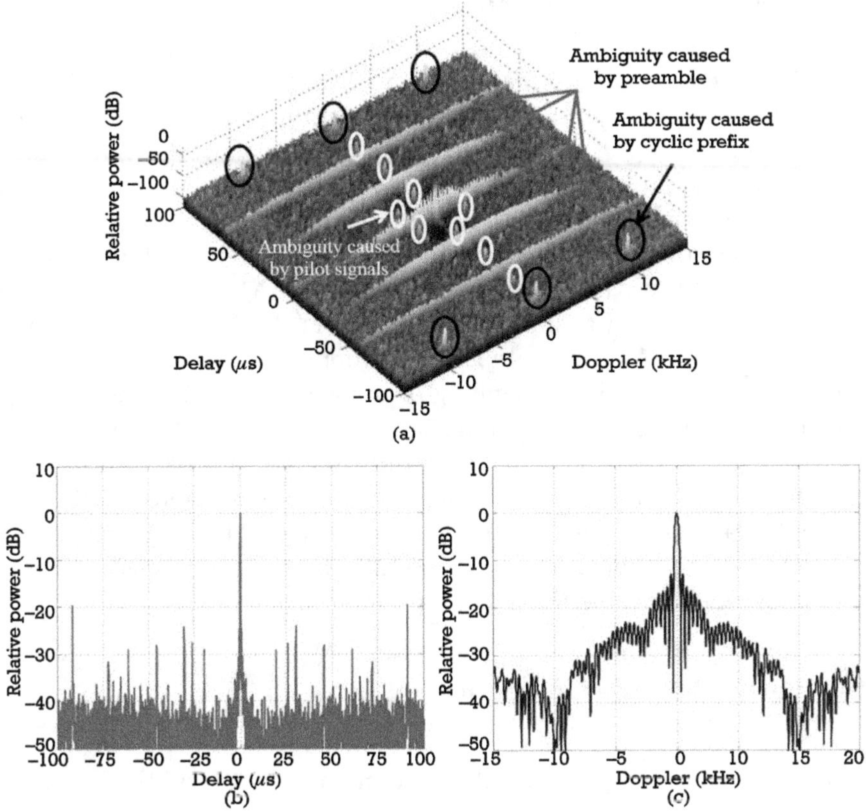

Figure 3.13 (a) Measured ambiguity function of one frame of WiMAX signal; (b) cut at zero Doppler; and (c) cut at zero range. (From: [37].)

measured ambiguity function of one frame of the WiMAX downlink signal.

The figure shows that the ambiguity function approximates to the thumbtack ideal with a range resolution of about 15m corresponding to the 10-MHz signal bandwidth and a Doppler resolution of about 330 Hz corresponding to the frame length. However, the figure also shows ridge-like ambiguities due to the preamble and point ambiguities due to the pilot signals and the cyclic prefix, as discussed in Section 3.3.3. Figure 3.13(b) shows the cut at zero Doppler (i.e., point target response in range), and Figure 3.13(c) shows the cut at zero range (i.e., point target response in Doppler).

References [35, 37] discuss mismatched filtering techniques that can be used to reduce the level of these ambiguities.

3.3.6 Digital Radio Mondiale

Yet another form of digital broadcast modulation is the Digital Radio Mondiale (DRM) modulation format, used for HF radio broadcasting. In DRM, the digitized audio stream is source coded using a combination of advanced audio coding (AAC) and spectral band replication (SBR) to reduce the data rate before time division multiplexing with two data streams (which are required for decoding at the receiver). A coded orthogonal frequency division multiplexing (COFDM) channel coding scheme is then applied, nominally with 200 subcarriers and a QAM mapping of these subcarriers is used to transmit the encoded data. This scheme is designed to combat channel fading, multipath and Doppler spread, allowing reception of data in the most demanding of propagation environments [38]. The ambiguity function, shown in Figure 3.14, has a well-defined peak and relatively uniform sidelobe level and structure, in the same way as the other digital modulation formats discussed in this chapter. In this example the range resolution of the signal, determined from the range cut, is 16 km and the Doppler resolution is 12.5 Hz, although longer integration times would give better Doppler resolution.

3.4 Vertical-Plane Coverage

The performance of a passive radar system depends not only on the waveform, but also on the coverage of the illuminating sources. The coverage

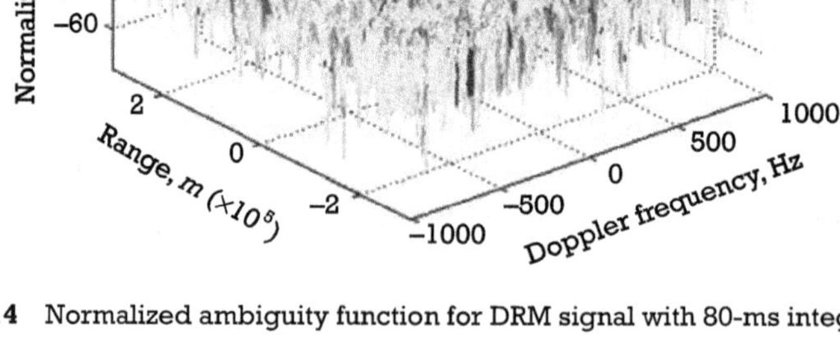

Figure 3.14 Normalized ambiguity function for DRM signal with 80-ms integration time [5].

of broadcast and communications transmitters will be optimized according to the required services, and the transmitters will frequently be sited on hilltops or on tall buildings. The horizontal-plane coverage is often omnidirectional, although for some cellphone base stations it may be arranged in 120° sectors. The vertical-plane coverage will usually be optimized so as to avoid wasting power above the horizontal, and in some cases the beams may be tilted downwards by a degree or so.

Examples have been published of measured vertical-plane field strength patterns of typical PBR transmitters (VHF FM and DVB-T) [39, 40]. These can be replotted in a more meaningful form (Figure 3.15) to show the reduction in power density illuminating a target as a function of the sine of the elevation angle. It can be seen that the antennas of the VHF FM transmitters have relatively high sidelobes, but the array of the DVB-T transmitter allows greater control of the radiation pattern and hence lower sidelobes.

Taking the radar equation for passive radar (Chapter 2) in the form:

$$\frac{S}{N}=\frac{P_t G_t G_r \lambda^2 \sigma_b G_p}{(4\pi)^3 R_T^2 R_R^2 kT_0 BFL} \tag{3.4}$$

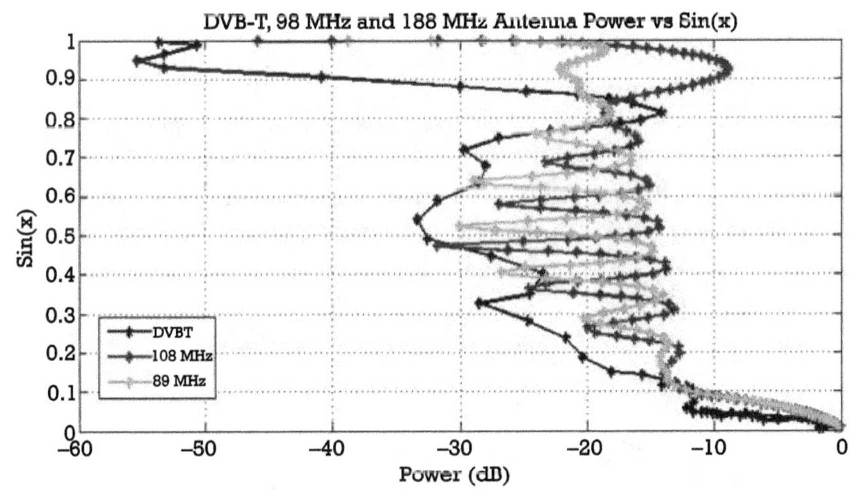

Figure 3.15 Measured vertical-plane radiation patterns of BBC VHF FM radio transmitter at 98 MHz, 108 MHz, and 8-bay DVB-T transmitter. The vertical scale is the sine of the elevation angle at the transmitter.

and rearranging:

$$R_{R\max} = \sqrt{\frac{P_t G_t G_r \lambda^2 \ \sigma_b \ G_p}{(4\pi)^3 \ R_T^2 \left(S/N\right)_{\min} \ kT_0 BFL}} \tag{3.5}$$

shows that for every 10-dB reduction in $P_t G_t$ the maximum detection range RR for a given target is reduced by a factor of 3.3 (Figure 3.16). Even at the peaks of the elevation-plane lobes the effect is significant, but in the nulls in between the lobes it is even more so.

3.5 Satellite-Borne Illuminators

The enormous growth at consumer scale in the past couple of decades in satellite broadcasting and navigation means that there are a large number of transmissions that may be used for passive radar purposes.

There is an important distinction here between satellites in geostationary orbit, in which case the illumination of the target scene is constant and continuous in time, and ones in low Earth orbit (LEO), in which case the illumination is brief (only a few seconds at most) but providing global or near-global coverage. As discussed at the beginning of this chapter, the important parameters of the signal are the power density at the target, and the waveform. For broadcast, communications, or navigation systems the power density at the Earth's surface will be such as to give a suitable signal-to-noise ratio with whatever receive antenna is used (such as a dish antenna for DBS TV or a handheld receiver for GNSS). However, for a space-borne radar the power density at the Earth's surface will be such as to give a detectable echo back at the radar receiver, and

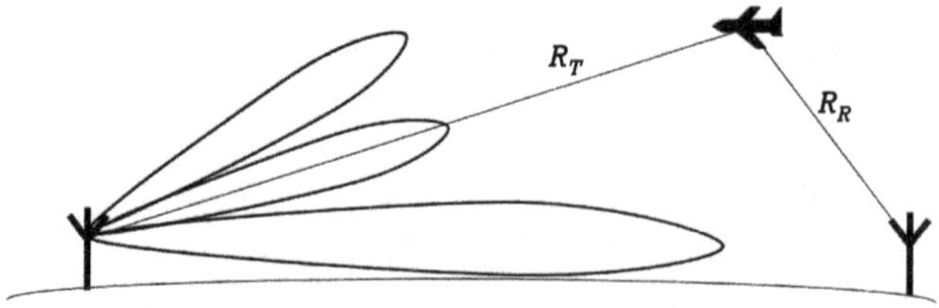

Figure 3.16 The effect on detection range of the elevation-plane pattern of the source can be substantial.

that power density will be substantially higher. Such signals are therefore much more suitable as passive radar illuminators.

Figure 3.17 summarizes the parameters of various satellite-borne illuminators.

3.5.1 Global Navigation Satellite System

Global Navigation Satellite System (GNSS) is an overall term for satellite navigation, and includes the original U.S. NAVSTAR Global Positioning System (GPS) and the Russian GLObal'naya NAvigatsionnaya Sputnikovaya Sistema (GLONASS). Further, the European GALILEO system, the Chinese BeiDou, and the Indian NAVIC systems are to set become fully operational by 2020. Although GPS was originally developed for military use and had both military and civil (lower resolution) codes, GNSS is now widely used for domestic and commercial vehicle navigation (satnav) and for surveying.

The signals are generally at L-band, modulated by pseudo-random noise (PRN) codes (CDMA for GPS, FDMA for GLONASS). Each system consists of a constellation of satellites at an orbit height of around 20,000

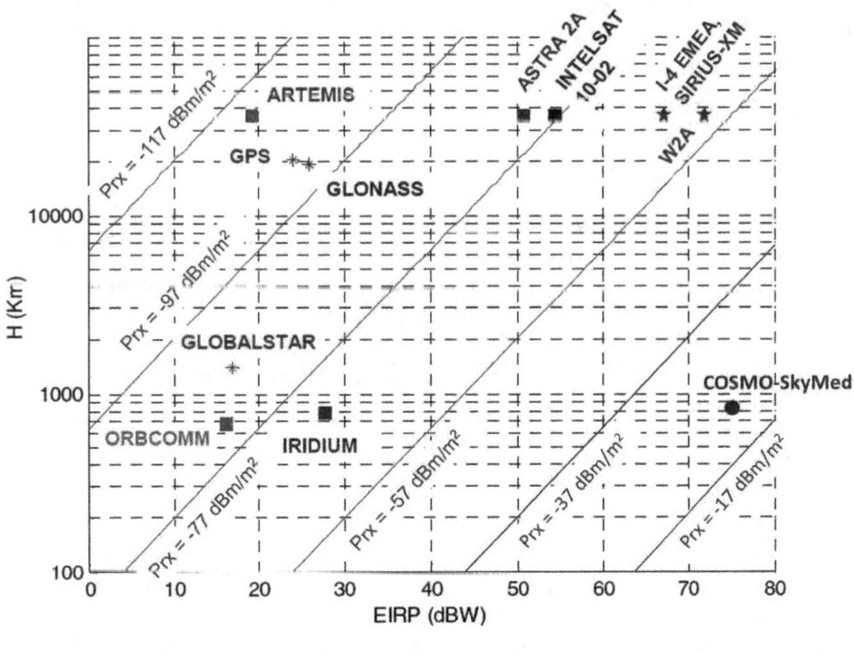

Figure 3.17 EIRPs, satellite heights and power densities (Prx) at the Earth's surface for various satellite-borne illuminators. (Adapted from: [41].)

km. Table 3.2 summarizes the orbit parameters of GPS, GLONASS, and GALILEO, and Figure 3.18 depicts the nominal orbit constellation for GPS.

3.5.2 Satellite TV

Geostationary satellites are positioned above the equator at a height of 35,786 km and orbit the Earth with a period of 24 hours. The original concept was published by the scientific writer Arthur C. Clarke in 1945 [43]. They therefore appear fixed and are used in applications such as satellite TV broadcasting and maritime communications (INMARSAT). For satellite TV, the antenna footprints are shaped so as to give coverage of specific land areas, which means that the coverage of ocean areas is poor.

Satellite TV signals are at Ku-band. Each satellite has typically 27 transponders, each with a bandwidth of between 27 and 50 MHz. The EIRP of the transmitters is typically +55 dBW, giving a power density at the Earth's surface of about −107 dBW/m^2 [44]. The DVB-S modulation format is similar to that of DVB-T.

3.5.3 INMARSAT

The INMARSAT system is used for maritime communications and currently consists of 12 satellites in geostationary orbit. The latest (INMARSAT-5) provide the high-speed Global Xpress broadband service, with three satellites:

- I-5 F1 EMEA, at 63° East;
- I-5 F2 Americans and Atlantic Ocean Region, at 55° West;

Table 3.2
Orbit Parameters of GPS, GLONASS, and GALILEO Systems

Parameter	GPS System	GLONASS System	GALILEO System
Number of active satellites	24	24	30
Number of orbital planes	6	3	3
Orbital inclination	55°	64.8°	56°
Orbit altitude	20,183 km	19,130 km	23,616 km
Orbital period	11 hours, 58 minutes, 0 seconds	11 hours, 15 minutes, 40 seconds	14 hours, 4 minutes
Absolute velocity	3,870 m/s	3950 m/s	3,720 m/s

Source: [42].

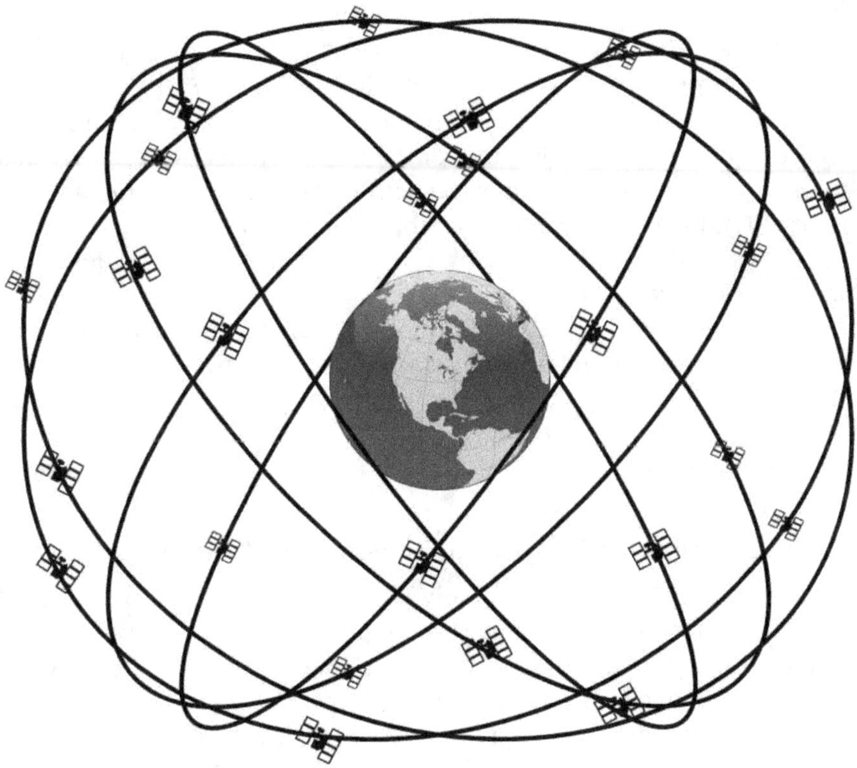

Figure 3.18 Constellation of GPS satellite orbits.

- I-5 F3 Pacific Ocean Region, at 179° East.

The Broadband Global Area Network (BGAN) service provides communications with ships and aircraft from satellites in geostationary orbit. The downlink is at L-band, with 630 channels each of 200-kHz bandwidth (432 kbps), 228 spot beams, and an EIRP of +67 dBW, which gives a power density of 3×10^{-10} W/m^2 at the Earth's surface.

Lyu et al. examined the form of the signal for passive radar purposes [45].The modulation is 16-QAM with a 0.25-root cosine shaping filter. By combining several adjacent channels, they demonstrate an ambiguity function with a peak sidelobe level of –18.7 dB. However, the range resolution is still rather coarse, and the power density is such that considerable integration gain will be required.

3.5.4 IRIDIUM

The IRIDIUM network consists of a constellation of 66 satellites in low Earth orbit (6 orbits of 11 satellites each) at a height of 781 km, providing global voice and data coverage to satellite phones and pagers. The system uses the 1,616–1,626.5-MHz band for both uplink and downlink, divided into 240 channels each with a bandwidth of 41.67 kHz. The modulation format is TDMA, with a frame length of 90 ms. Each frame begins with a 20.32-ms simplex slot, followed by four uplink slots and four downlink slots, each of length 8.28 ms.

Lyu et al. also examined the form of this signal for passive radar purposes [46]. By combining several adjacent channels they demonstrate a peak sidelobe level of −18.7 dB. However, the range resolution is still rather coarse, and the power density is such that considerable integration gain will be required.

3.5.5 Low Earth Orbit Radar Remote-Sensing Satellites

Since the mid-1970s, satellites in low Earth orbit have been widely used for geophysical remote sensing. These typically carry a suite of optical, infrared, and radar instruments, and the signals from the radars may be exploited as passive radar sources. The principal type of radar used in this way is the synthetic aperture radar (SAR), although in principle other types of satellite-borne radar (radar altimeter, scatterometer) could also be used.

Such satellites are placed in near polar orbits giving almost global coverage with an orbit period of typically 100 minutes and an orbit pattern that repeats after a fixed interval, typically between 3 and 30 days.

Satellite SAR signals are typically chirped pulses of bandwidth from tens to hundreds of megahertz, at frequencies from L-band to X-band. The pulse repetition frequency is of the order of kilohertz (this is chosen to avoid ambiguities in range and in Doppler). They are radiated at high power from large antennas pointing to one side of the subsatellite track (Figure 3.19), and most recent SAR designs include multiple modes that incorporate elevation-plane scanning (SCANSAR) to give increased swath width and/or azimuth-plane scanning to give increased resolution (spotlight mode). Many also include polarimetric modes, radiating pulses of alternate horizontal and vertical polarization, and some (TanDEM-X, COSMO-SkyMed ...) consist of constellations of multiple satellites, allowing for interferometric experiments [47].

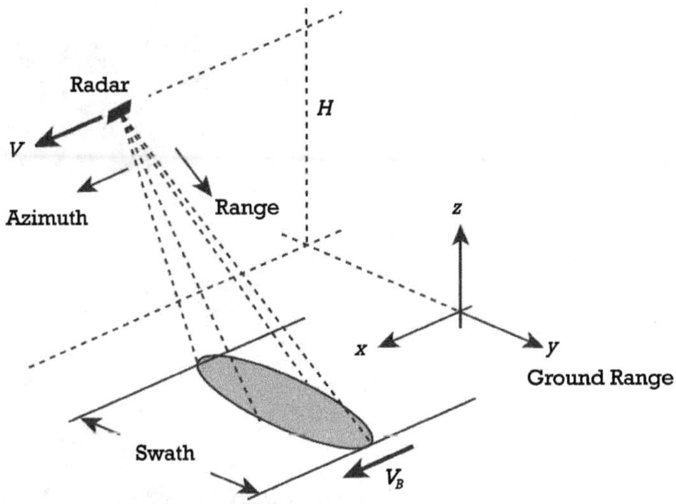

Figure 3.19 Satellite SAR geometry. The antenna looks to the side of the subsatellite track, illuminating a swath of the imaged target scene.

Figure 3.20 shows an example of direct reception of the signal from a satellite-borne SAR (in this case the ASAR instrument carried by the European Space Agency's ENVISAT satellite). The signal-to-noise ratio here is 25 dB, even with a small horn receive antenna. The figure shows the sequence of linear-FM chirp pulses both in the time domain and in the frequency domain.

It is straightforward to obtain a clean version of the direct signal, using an upward-pointing antenna and a separate receiver, which is how the results in Figure 3.20 were obtained.

3.6 Radar Illuminators

The final class of illuminator source is radar transmitters, in which case the technique is known as hitchhiking. Such sources may be cooperative, in which case the location, frequency, pulse length, PRF, and antenna scan pattern of the transmitter are known and can be optimized or noncooperative, in which case they are not. The radar transmitter may be terrestrial, airborne or shipborne.

Just as with the satellite-borne radar illuminators considered in the previous section, the signal will be very suitable for radar operation, typically consisting of pulses at a regular PRF. The ambiguity function should

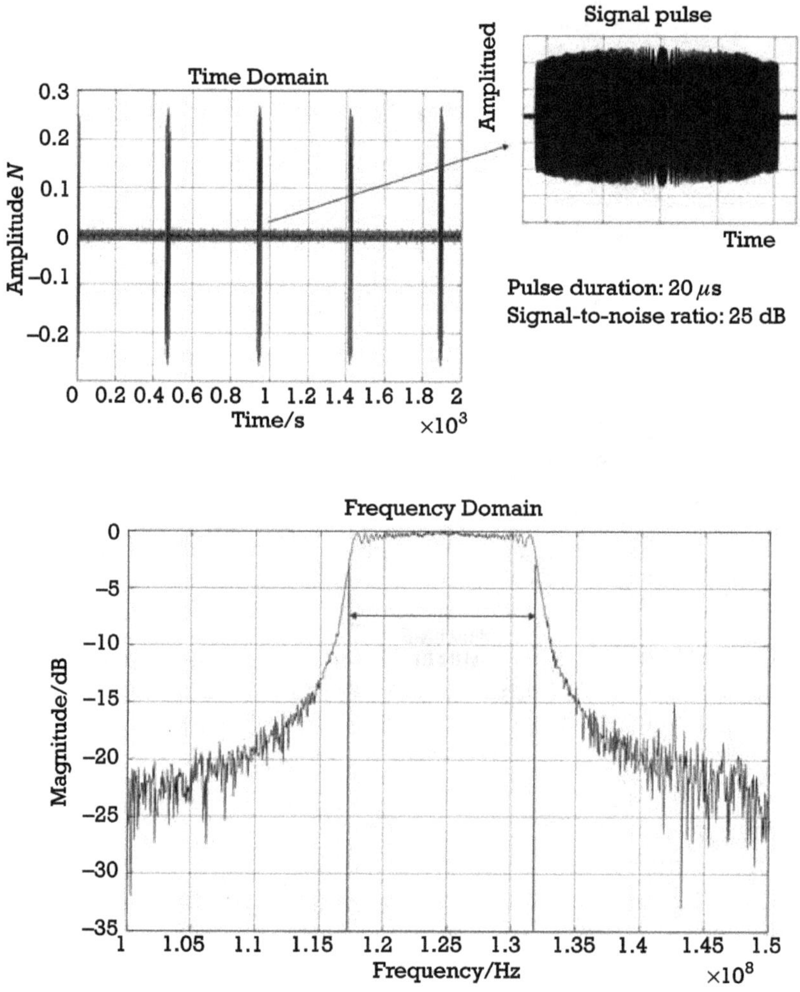

Figure 3.20 Direct radar pulses received from an overpass of the European Space Agency's ENVISAT SAR. The pulses are linear FM chirps, with a duration of 20 μs and a bandwidth of 15 MHz [48].

be very favorable. Furthermore, the power density at the target will be high.

The passive radar receiver needs to be able to synchronize to the instant of pulse transmission and the pointing direction of the transmit antenna. In almost all cases, the radar transmitter will scan in azimuth (Figure 3.21), and in general the direct signal from the radar will not be detectable over the full range of transmitter azimuth scan angles.

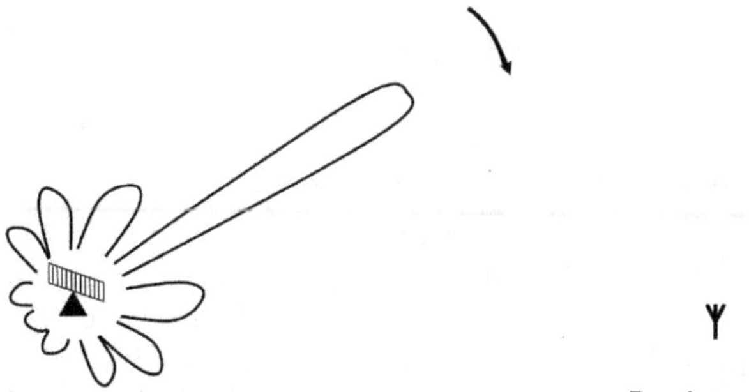

Figure 3.21 Passive radar hitchhiking based on a scanning radar transmitter.

However, it is possible to use flywheel clocks at the passive radar receiver for the PRF and for the transmit antenna pointing direction, which are resynchronized each time the antenna beam sweeps past [49].

If the passive radar is to use a directional antenna, which would be desirable both to give angular discrimination and to provide gain, its beam must follow the transmitted pulse through space so that it is pointing in the direction from which target echoes would come at that instant. This is called *pulse chasing* and is depicted in Figure 3.22 [50]. The instantaneous receive beam direction is given by:

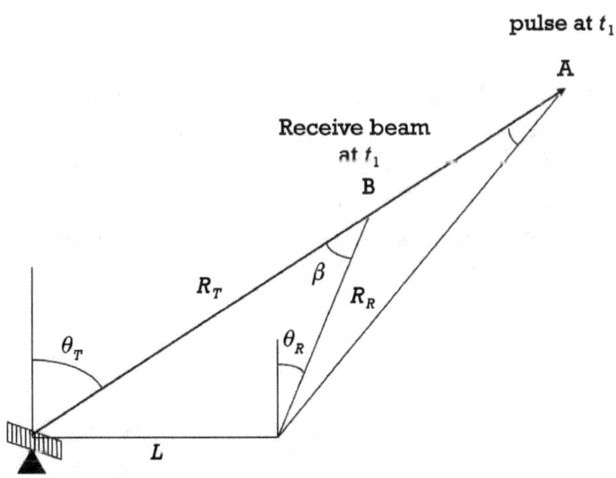

Figure 3.22 Pulse chasing: the receiver beam must scan at a very rapid and nonlinear rate.

$$\theta_R = \theta_T - 2\tan^{-1}\left(\frac{L\cos\theta_T}{R_T + R_R - \sin\theta_T}\right) \tag{3.6}$$

where the symbols are defined in Figure 3.22. It is important to realize that the receive beam does not point at the instantaneous position of the target, but in the direction from which the target echo comes at the instant it arrives at the receiver. Thus, at time t_1 after the instant of transmission, the pulse has reached point A, but the receive beam must point towards point B to take account of the propagation time from point B to the receiver. It can also be appreciated that the scan rate of the receive beam is highly nonlinear, being most rapid when the beam is perpendicular to the direction of propagation of the pulse.

Such a rapid and nonlinear scan rate could not be achieved by mechanical scanning, so the receive antenna would have to be an electronically scanned array, which means that one of the most significant potential advantages of passive radar (simplicity and low cost) is lost.

Finally, within the class of radar illuminators, mention should be made of HF over-the-horizon radars (OTHRs). These operate in the HF band (2–30 MHz) and achieve very long ranges by reflection of their signals from the ionosphere [51]. The frequency is selected adaptively to achieve the desired range according to the reflective properties of the ionosphere which vary according to the time of day, time of year, and the sunspot cycle, and the signals are coded pulses. These, too, are usable as sources for bistatic hitchhiking.

In summary, radar transmitters are highly suitable as illuminators for passive radar, since the signals are already optimized for radar operation and the power density at the target is high. However, if the receiver is to use a directional antenna it is necessary to use pulse chasing, which introduces significant complication and cost. References [49, 52] described two systems using omnidirectional receive antennas, showing that for some applications at least, this can be avoided.

3.7 Summary

This chapter has reviewed the properties of illuminator sources that may be used for passive radar purposes. These properties are fundamental in defining the performance of such systems, and so they need to be understood and quantified in selecting the right source to use. The key parameters are the power density at the target, the nature of the waveform, and the coverage. There is a wide range of signals that may be used. Just as with a monostatic radar, the ambiguity function provides an elegant way of showing the waveform properties, such as resolution, ambiguities, and sidelobe structure, all in range and in Doppler.

Emphasis has been placed in this chapter on digital modulation formats, since these are increasingly replacing analog modulation in communications, broadcast and radionavigation. With analog signals, the ambiguity function will in general be time-varying and will depend on the nature of the program content. Digital signals tend to be more noise-like, and the ambiguity function does not depend on the program content and is not time-varying. Nevertheless, periodic features in digital modulation formats will result in corresponding periodic features in the ambiguity function. Several different modulation formats have been described in this chapter. Modulation formats based on OFDM are widely and increasingly used, since they provide a means of suppressing the effects of multipath.

We have considered the effect of the elevation-plane coverage of terrestrial illuminators. In practice, this may prove to be a significant limitation, particularly in the detection and tracking of air targets at high elevation angles.

A number of satellite-borne illuminators have been described, including LEO remote-sensing SARs, GNSS, and geostationary sources including satellite TV and INMARSAT. LEO remote-sensing SAR signals have the advantage that the signals are already optimized for radar purposes and the power density at the target is high, but the illumination is brief. Signals from satellites in geostationary orbit are significantly weaker, but give continuous illumination, and hence potential (with suitable targets) for integration gain.

Finally, we have considered conventional radar illuminators, for which the technique is known as *hitchhiking*. Here, too, the signals are already optimized for radar purposes and the power density at the target is high. However, if the receiver is to use a directional antenna, pulse chasing is necessary, which introduces significant extra complexity.

References

[1] Woodward, P. M. *Probability and Information Theory, with Applications to Radar*, London, U.K.: Pergamon Press, 1953; reprinted, Dedham, MA: Artech House, 1980.

[2] Tsao, T., et al., "Ambiguity Function for a Bistatic Radar," *IEEE Trans. on Aerospace and Electronic Systems*, Vol. 33, No. 3, July 1997, pp. 1041–1051.

[3] Ringer, M. A., and G. J. Frazer, "Waveform Analysis of Transmissions of Opportunity For Passive Radar," *Proc. ISSPA'99*, Brisbane, August 22–25, 1999, pp. 511–514.

[4] Griffiths, H. D., et al., "Measurement and Analysis of Ambiguity Functions of Off-Air Signals for Passive Coherent Location," *Electronics Letters*, Vol. 39, No. 13, June 26, 2003, pp. 1005–1007.

[5] Thomas, J. M., H. D. Griffiths, and C. J. Baker, "Ambiguity Function Analysis of Digital Radio Mondiale Signals for HF Passive Bistatic Radar," *Electronics Letters*, Vol. 42, No. 25, December 7, 2006, pp. 1482–1483.

[6] Griffiths, H. D., and C. J. Baker, "Passive Bistatic Radar," Ch. 11 in *Principles of Modern Radar, Vol. 3*, W. Melvin, (ed.), Raleigh, NC: SciTech Publishing, 2012.

[7] Torre, A., and P. Capece, "COSMO-SkyMed: The Advanced SAR Instrument," *5th International Conference on Recent Advances in Space Technologies (RAST)*, Istanbul, June 9–11, 2011.

[8] Tasdelen, A. S., and H. Köymen, "Range Resolution Improvement in Passive Coherent Location Radar Systems Using Multiple FM Radio Channels," *IET Forum on Radar and Sonar*, London, U.K., November 2006, pp. 23–31.

[9] Bongioanni, C., F. Colone, and P. Lombardo, "Performance Analysis of a Multi-Frequency FM Based Passive Bistatic Radar," *IEEE Radar Conference*, Rome, Italy, May 26–30, 2008.

[10] Olsen, K. E., "Investigation of Bandwidth Utilisation Methods to Optimise Performance in Passive Bistatic Radar," Ph.D. thesis, University College London, 2011.

[11] Olsen, K. E., and K. Woodbridge, "Performance of a Multiband Passive Bistatic Radar Processing Scheme – Part I," *IEEE AES Magazine*, Vol. 27, No. 10, October 2012, pp. 17–25.

[12] Olsen, K. E., and K. Woodbridge, "Performance of a Multiband Passive Bistatic Radar Processing Scheme – Part II," *IEEE AES Magazine*, Vol. 27, No. 11, November 2012, pp. 4–14.

[13] Zaimbashi, A., "Multiband FM-Based Passive Bistatic Radar: Target Range Resolution Improvement," *IET Radar, Sonar and Navigation*, Vol. 10, No. 1, January 2016, pp. 174–185.

[14] Christensen, J. M., and K.E. Olsen, "Multiband Passive Bistatic Radar Coherent Range and Doppler Walk Compensation," *IEEE Int. Conference RADAR 2015*, Arlington VA, May 11–14, 2015, pp. 123–126.

[15] Saini, R., and M. Chemiakov, "DTV Signal Ambiguity Function Analysis for Radar Application," *IEE Proc. Radar, Sonar and Navigation*, Vol. 152, No. 3, 2005, pp. 133–142.

[16] Rappaport, T., *Wireless Communications: Principles and Practice*, 2nd ed., Upper Saddle River, NJ: Prentice-Hall, 2001.

[17] Tan, D. K. P., et al., "Passive Radar Using The Global System for Mobile Communication Signal: Theory, Implementation and Measurements," *IEE Proc. Radar, Sonar and Navigation*, Vol. 152, No. 3, June 2005, pp. 116–123.

[18] *3GPP standard, LTE; Evolved Universal Terrestrial Radio Access (E- UTRA);* Base Station (BS) radio transmission and reception (3GPP TS 36.104 version 10.2.0 Release 10).

[19] http://media.ofcom.org.uk/news/2013/winners-of-the-4g-mobile-auction/, accessed September 5, 2016.

[20] Dahlman, E., S. Parkvall, and J. Sköld, *4G-LTE/LTE-Advanced for Mobile Broadband*, New York: Elsevier, 2013.

[21] LTE: Evolved Universal Terrestrial Radio Access (E-UTRA), 3GPP standard document, ETSI TS – B6.211 V10.0.0, 2011.

[22] http://uk.mathworks.com/ , accessed October 23, 2016.

[23] Sutton, P. D., K. E. Nolan, and L. E. Doyle, "Cyclostationary Signatures for Rendezvous in OFDM-Based Dynamic Spectrum Access Networks," *IEEE DySPAN 2007*, Dublin, Ireland, April 17–20, 2007.

[24] Alhabashna, A., et al., "Cyclostationarity-Based Detection of LTE OFDM Signals for Cognitive Radio Systems," *IEEE Globecom Conference 2010*, Miami FL, December 6–10, 2010.

[25] Evers, A., and J. Jackson, "Analysis of an LTE Waveform for Radar Applications," *IEEE Radar Conference 2014*, Cincinnati, OH, May 19–23, 2014.

[26] Griffiths, H. D., I. Darwazeh, and M. I. Inggs, "Waveform Design for Commensal Radar," *IEEE Int. Radar Conference 2015*, Arlington VA, May 11–14, 2015, pp. 1456–1460.

[27] Harms, H. A., L. M. Davis, and J. E. Palmer, "Understanding the Signal Structure in DVB-T Signals for Passive Radar Detection," *IEEE Int. Radar Conference 2010*, Washington, D.C., May 10–14, 2010, pp. 532–537.

[28] Palmer, J. E., et al., "DVB-T Passive Radar Signal Processing," *IEEE Trans. on Signal Processing*, Vol. 61, No. 8, April 2013, pp. 2116–2126.

[29] *Digital Video Broadcasting (DVB): Framing Structure, Channel Coding and Modulation for Digital Terrestrial Television (DVB-T)*, 1st ed., European Telecommunications Standards Institute, March 1997.

[30] Berger, C. R., et al., "Signal Processing for Passive Radar Using OFDM Waveforms," *IEEE J. Selected Topics in Signal Processing*, Vol. 4, No. 1, February 2010, pp. 226–238.

[31] IEEE Standards: *Information Technology. Part 11: Wireless LAN Medium Access Control (MAC) and Physical Layer (PHY) Specifications* (IEEE Std. 802.11TM-1999). *Supplements and Amendments* (IEEE Stds 802.11aTM-1999, 802.11bTM-1999, 802.11bTM-1999/Cor 1-2001, and 802.11gTM-2003).

[32] *IEEE Standard for Local and Metropolitan Area Networks Part 16: Air Interface for Fixed Broadband Wireless Access Systems*, Rev. IEEE Standard 802.16-2004, Oct. 2004, (revision of IEEE Standard 802.16-2001).

[33] *IEEE Standard for Local and Metropolitan Area Networks Part 16: Air Interface for Fixed and Mobile Broadband Wireless Access Systems Amendment 2: Physical and Medium Access Control Layers for Combined Fixed and Mobile Operation in Licensed Bands and Corrigendum 1*, Rev. IEEE Standard 802.16e-2005 and IEEE Standard 802.16-2004/Cor 1-2005, February 2006 (amendment and corrigendum to IEEE Standard 802.16-2004).

[34] Colone, F., et al., "Ambiguity Function Analysis of Wireless LAN Transmissions for Passive Radar," *IEEE Trans. Aerospace and Electronic Systems*, Vol. 47, No. 1, January 2011, pp. 240–264.

[35] Colone, F., P. Falcone, and P. Lombardo, "Ambiguity Function Analysis of WiMAX Transmissions for Passive Radar," *IEEE Int. Radar Conf.*, Arlington, VA, May 10–14, 2010, pp. 689–694.

[36] Wang, Q., Y. Lu, and C. Hou, "Evaluation of WiMAX Transmission for Passive Radar Applications," *Microwave and Optical Technology Letters*, Vol. 52, No. 7, 2010, pp. 1507–1509,.

[37] Higgins, T., T. Webster, and E. L. Mokole, "Passive Multistatic Radar Experiment Using WiMAX Signals of Opportunity Part 1: Signal Processing," *IET Radar, Sonar and Navigation*, Vol. 10, No. 2, February 2016, pp. 238–247.

[38] Hoffman, F., C. Hansen, and W. Shäfer, "Digital Radio Mondiale (DRM) Digital Sound Broadcasting in the AM bands," *IEEE Trans. on Broadcast.*, Vol. 49, No. 3, 2003, pp. 319–328.

[39] Millard, G. H., *The Introduction of Mixed-Polarization for VHF Sound Broadcasting: the Wrotham Installation,* Research Department Engineering Division, British Broadcasting Corporation, BBC RD 1982/17, September 1982.

[40] O'Hagan, D. W., et al., "A Multi-Frequency Hybrid Passive Radar Concept for Medium Range Air Surveillance," *IEEE AES Magazine,* Vol. 27, No. 10, October 2012, pp. 6–15.

[41] Cristallini, D., et al., "Space-Based Passive Radar Enabled by the New Generation of Geostationary Broadcast Satellites," *IEEE Aerospace Conf.,* Big Sky MT, March 2010.

[42] Eissfeller, B., et al., "Performance of GPS, GLONASS and Galileo," *Photogrammetric Week '07,* D. Fritsch, (ed.) Wichmann Verlag, Heidelberg, 2007, pp. 185–199.

[43] Clarke, A. C., "Extra-Terrestrial Relays," *Wireless World,* October 1945, pp. 305–308.

[44] Griffiths, H. D., et al., "Bistatic Radar Using Satellite-Borne Illuminators of Opportunity," *IEE Int. Radar Conference RADAR-92,* Brighton; IEE Conf. Publ. No. 365, October 12–13, 1992, pp. 276–279.

[45] Lyu, X., et al., "Ambiguity Function of Iridium Signal for Radar Application," *Electronics Letters,* Vol. 52, No. 19, September 15, 2016, pp. 1631–1633.

[46] Lyu, X., et al., "Ambiguity Function of Inmarsat BGAN Signal for Radar Application," *Electronics Letters,* Vol. 52, No. 18, September 2, 2016, pp. 1557–1559.

[47] Griffiths, H. D., C. J. Baker, and D. Adamy, *Stimson's Introduction to Airborne Radar,* 3rd ed., Ch. 35: 'SAR System Design," Raleigh, NC: Scitech Publishing, May 2014.

[48] Whitewood, A., C. J. Baker, and H. D. Griffiths, "Bistatic Radar Using a Spaceborne Illuminator," *IET Int. Radar Conference RADAR 2007,* Edinburgh, October 15–18, 2007.

[49] Schoenenberger, J. G., and J. R. Forrest, "Principles of Independent Receivers for Use with Co-Operative Radar Transmitters," *The Radio and Electronic Engineer,* Vol. 52, No. 2, February 1982, pp. 93–101.

[50] Jackson, M. C., "The Geometry of Bistatic Radar Systems," *IEE Proc.,* Vol. 133, Pt. F, No. 7, December 1986, pp. 604–612.

[51] Headrick, J. M., and J. F. Thomason, "Applications of High-Frequency Radar," *Radio Science,* Vol. 33, No. 4, July-August 1998, pp. 1045–1054.

[52] Hawkins, J. M., "An Opportunistic Bistatic Radar," *IEE Int. Radar Conference RADAR 97,* Edinburgh, October 14–16, 1997, pp. 318–322.

CHAPTER

4

Contents

Direct Signal Suppression

4.1 Introduction

Chapter 2 briefly introduced the concept of direct signal interference. You will recall that the direct signal is the part of the transmitted signal that arrives directly at the receiver without having been reflected by a target. At its simplest, it is the signal that travels along the baseline of any bistatic pair in a passive radar network (Figure 4.1). It is always present in passive radar systems, especially those that exploit illuminators such as very high frequency (VHF) radio and ultrahigh frequency television (UHF TV) stations that tend to transmit in all azimuth directions. The direct signal is only subject to a one-way propagation loss and hence is attenuated by $1/L^2$, where L is the baseline distance between the transmitter and the receiver (Figure 4.1). Because the signal strength at the receiver is only attenuated by $1/L^2$ and as the baseline is always less than the

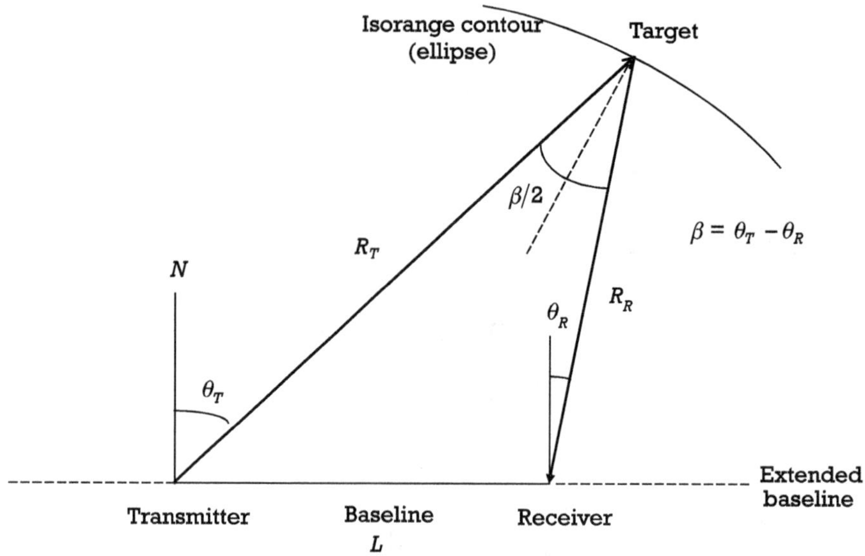

Figure 4.1 The bistatic radar geometry. The target velocity is v, making an angle δ with the bisector of the bistatic angle β.

bistatic range ($R_T + R_R$), the directly received signal can be very strong compared to the weak target echoes.

A passive radar system exploits the directly received signal to provide a timing reference against which the signal reflected from the target can be compared and hence receiver range to the target evaluated. This is achieved via a cross-correlation of the directly and indirectly received signals. However, the direct signal will also leak into the surveillance channel antenna being used to detect targets. The resulting direct signal interference has to be reduced to a level that, in the ideal case, is below that of receiver noise so that it avoids any reduction in the maximum possible detection range of a target of given radar cross-section.

Before quantifying the level of directly received signal power, the directly received signal is considered in a little more detail. This is illustrated in simple schematic form in Figure 4.2. Here, it can be seen that in addition to the directly received signal and the indirect signal from a target there are also reflections received in the surveillance channel. These reflections can occur from buildings and natural objects such as trees. In fact, it can often be the case that a building close to the passive radar surveillance antenna acts as a strong reflector such that the reflections can be at a power level that is still very high but occur at an angle other than

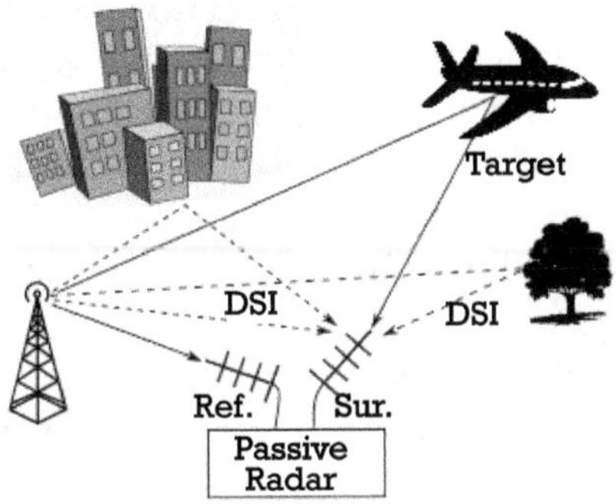

Figure 4.2 Illustration of passive radar direct signal interference scenario.

that of the direction of the illuminator being exploited. As a consequence, these replicas of the direct signal, appearing at different angles to the location of the illuminator, can greatly complicate the suppression of all possible unwanted signal in the surveillance channel. Collectively, these signals limit the detection range of a passive radar system unless steps are taken to minimize their presence. It also places great importance on a selecting a suitable site for the location of the passive radar receiver, along with other considerations such as bistatic geometry, unrestricted field of regard, and so forth. It should also not be forgotten that the directly received signals, including that coming from the direction of the illuminator of opportunity, can also be subject to multipath via ground bounce as well as reflections from nearby objects.

In this chapter the form of the directly received signal is examined, as it appears in the surveillance channel, together with methods used to reduce its signal strength to enable satisfactory target detection ranges to be achieved. For clarity, we restrict ourselves here to the case in which a single directly received signal enters the surveillance channel after travelling directly along the passive radar system baseline.

4.2 Direct Signal Interference Power Levels

A simple expression can be written down for the amount of direct signal suppression required by first calculating the ratio of the indirect received signal to the direct signal, and second, being required, as a design goal, for the level of direct signal suppression to be such that the highest level of interference tolerable is equivalent to that for single pulse-like detection as limited by receiver noise. Note that there is no benefit from integration as the direct signal that leaks into the surveillance channel will also integrate up. Indeed, in practice, this may lead to an even more stringent requirement for direct signal suppression needing. It is also made more complex, as stated above, if there are multiple local copies of the direct signal that are also received in the surveillance channel. In simple terms, however, this representation will place the direct leakage signal and the noise floor in the receiver at the same level. Consequently, it has the attractive feature of providing equivalent performance to single-pulse detection and also maximising detection range for a target of a given radar cross-section.

To gain a feel for the amount of direct signal, first, the power received from the target can be computed using the bistatic radar equation introduced in Chapter 2 in the form shown in:

$$P_{tar} = \frac{P_t^{av} G_t^{tar} G_r^{m} \lambda^2 \sigma_b}{(4\pi)^3 R_t^2 R_r^2 L_b L_s} \tag{4.1}$$

where P_t^{av} is the average transmitted power of illuminator of opportunity, G_t^{tar} is the gain of transmit antenna towards the target, G_r^m is the surveillance antenna mainbeam gain towards target, λ is the wavelength, σ_b is the bistatic radar cross-section, R_t is the transmit range (transmitter to target), R_r is the receiver range (receiver to target), L_b is the propagation losses along the bistatic path, and L_s is the system losses (e.g., antenna mismatch, cabling, and so forth).

The direct signal interference, in its most general form, consists not only of the direct path breakthrough into the sidelobes of the surveillance antenna, but also strong multipath and clutter from the environment around the receiver at various time delays and azimuth angles. These scattered components typically sit at levels between the direct path breakthrough and thermal noise floor. However, assuming the direct path leakage is the dominant direct signal interference term, for simplicity, the direct signal interference power at the antenna can be approximated using the Friis transmission equation (4.2). The direct signal interference received power in the surveillance channel can be formulated as:

$$P_{dsi} = \frac{P_t^{av} G_t^{r} G_r^{s} \lambda^2}{(4\pi)^2 R_L^2 L_d L_s} \tag{4.2}$$

where G_t^r is the gain of transmit antenna towards the receiver, G_r^s is the surveillance antenna sidelobe gain towards transmitter, and is the propagation losses along the direct path.

If we assume $L_b = L_d$, $G_t^{tar} = G_t^r$ and $0.1G_r^m = G_t^s$ and make the further simplification that $R_L = R_T = R_R$ (a particular but not wholly unusual case), the ratio of (4.1) to (4.2) is subsequently given by

$$P_{tar} / P_{dsi} = \frac{\sigma_b}{(4\pi)\ R_L^2} \tag{4.3}$$

Thus, for a target of radar cross-section equal to 1 m^2, the direct signal interference is inversely proportional to the square of the baseline distance. Hence, if a baseline of 10 km were to be used, the ratio predicted by (4.3) would be 91 dB. Indeed, it is not at all unusual for this to exceed 10 0 dB, a very large factor to reduce and hence a challenging area of passive radar design that immediately impacts performance. Clearly, for other geometries, the ratio of (4.1) to (4.2) needs to be used in its complete form.

In this simple example, it has been assumed that the direct signal enters the surveillance channel via sidelobes at a level set to 10 dB below those of the main lobe. For many of today's systems with relatively modest array antennas, this is not an unrealistic assumption. Note, however, that array antennas themselves allow adaptive beamforming, which, in turn, provides a means of reducing the direct signal from even entering the surveillance channel. Hence, this would be a first stage in a strategy to minimize the direct signal interference. The combination of high gain antennas and adaptive beamforming also enables multiple simultaneous transmissions to be exploited. For systems with adaptive array antennas capable of nulling the direct path from the transmitter, or for systems with significant analog cancellation abilities, the direct path breakthrough of (4.2) may subsequently be weaker than a strong close in clutter response but with appreciable delay relative to the direct path. Scatterers with time delays greater than the inverse of the signal bandwidth cannot be removed with simple analog suppression methods, such as the Howells-Applebaum loop [1]. In such scenarios, a more appropriate representation for the direct signal interference power can be obtained by modifying (4.1) to sum over a number of clutter discretes, using the appropriate range, gain, and radar cross-section terms.

It should also be noted that the receiving surveillance antenna could be cross-polarized to further decrease the effective gain towards the transmitter, assuming that the target will have a significant cross polarization component as to not diminish significantly. In practice, the direct signal interference power level due to these complex interactions is extraordinarily difficult to predict and must be measured in situ for accurate estimates. In general, the objective for suppression of the directly received signal is to reduce it to a level below that of receiver noise. As receiver noise is a fundamental limit on detection range for a given set of radar parameters and target radar cross-section, this represents the ideal case. If the level of residual interference is above receiver noise, it will have an impact on detection range. This can be represented, in simple form, by an additional loss term in the radar range equation, (4.1).

4.3 Direct Signal Suppression

There are several techniques that may be used to suppress the direct signal appearing in the surveillance channel where it is at a level that reduces the maximum detection range of a target of a given radar cross section. These include physical shielding, Fourier processing, adaptive beamforming, and adaptive filtering. We consider the relative merits of each of these in turn.

Physical shielding can include anything from the direct channel being physically separated from the surveillance channel using a building or similar structure. The goal here is to have the building acting as a shield so that it stops as much of the direct signal as possible from reaching the surveillance antenna. This depends on the siting and construction of the building and cannot be relied upon. In the same way, geographical features, such as mountains, can be used and this has been done with great success by the research group at the University of Washington and is discussed in Chapter 7. In addition, the use of radar absorbing material (RAM) or radar fences can be employed to reduce the direct signal interference and this approach can be effective against sources arising from local features that cause the direct signal to enter from a range of different angles. These methods of physically shielding the surveillance channel from direct signal interference can be used individually or collectively to assist in achieving acceptable levels of suppression. However, for the relatively low frequencies (VHF and UHF) that tend to dominate the design of most passive radar systems, their effectiveness can be very limited as the signals at these frequencies lack directionality. As a consequence, they are unlikely to approach the necessary 100 dB predicted above and will need to form part of a more comprehensive set of methods to achieve the best possible suppression.

With regard to *Fourier processing*, the majority of passive radar systems are designed to detect aircraft and other air targets. In other words, they are detecting moving targets that, for the most part, will have a significant radial velocity or Doppler as seen in the surveillance channel. The direct signal, however, has no such motion component and hence simple Fourier processing offers a further degree of suppression of the direct signal in the surveillance channel. For example, a VHF signal with a bandwidth of 10 kHz appropriately sampled would lead to a fast Fourier transform (FFT) with around 10,000 data values per second and hence has a suppression factor of around 40 dB. A UHF signal with a bandwidth of 10 MHz would correspondingly have a suppression factor of about 70 dB. However, it should be noted that significant sidelobe leakage can occur that is due to the natural sidelobes of a bandlimited signal and spectral broadening induced by local clutter sources that exhibit internal motion characteristics (such as trees being blown by the wind). As a consequence, suppression levels, especially those close to zero Doppler, will not be as high as they will be at higher Doppler values.

With regard to *adaptive antenna nulling*, if an array antenna is used for the surveillance channel rather adaptive beamforming can be used to desensitize the receive channel in the direction of the transmitter (and others if required) through the formation of a null in that direction. This allows the directional gain of the antenna to provide suppression via control of the sidelobes. If a fully digital antenna is employed, then adaptive beamforming can be used to minimize sensitivity in the direction of the location of the directly received signal. If external noise such as multipath is present, then multiple nulls may have to be formed. If the external noise environment is nonstationary, the cancellation will need to be continuously adapting, with a suitably rapid response time. The number of degrees of freedom, and hence the number of antenna elements and receiver channels, must be greater than the number of signal components to be suppressed. The antenna pattern factor, the transmitter and receiver locations, and the target trajectory for a given scenario will lead to blind zones. These are caused either by a loss of line of sight among the transmitter, target, and receiver or when the target traverses the bistatic baseline between the transmitter and receiver. Most array antennas used in passive radar contain relatively few elements, typically 8 to 12. This reduces the degrees of freedom available, the achievable level of nulling and the number of nulls. It also restricts the gain in the direction of targets. It may be that in the future there will be passive arrays with larger antennas and these have the obvious advantages of higher gain, directionality as well as more degrees of design freedom, but come at the cost of extra size and complexity.

With regard to *adaptive filtering*, a general model for adaptive filtering, but one tailored to the direct signal interference suppression problem, is shown in Figure 4.3. The reference waveform, sampled at an interval, T_s, such that $t = nT_S$, is passed through a finite impulse response (FIR) filter with an impulse response $h[i]$, representing the direct path and clutter delay and complex scattering coefficients comprising the directly received signal in the surveillance channel.

$$s_{dsi}[n] = \sum_{i=0}^{M-1} h^*[i] s_r[n-i] \tag{4.4}$$

M is the number of discrete time delay coefficients required to properly model the various directly received signal components and $M = t_{\max}/T_s$. The various discrete samples of $h[i]$ represent a continuum of clutter responses, and as such there is no absolute guarantee that clutter responses will arrive at an integer number of sample delays. In most cases, a response for a given clutter discrete will be spread across neighboring coefficients of h, with diminished amplitude and a modified phase.

The direct signal interference removal process is one that first estimates the unknown clutter and direct path coefficients, $\hat{h}[i]$, and then convolves the result with the reference channel waveform to estimate the direct signal $s_{dsi}[n]$. This output is subtracted from to ideally leave just the target responses and thermal noise in surveillance signal. The final output after these operations is represented by $s_c[n]$, a supposedly direct

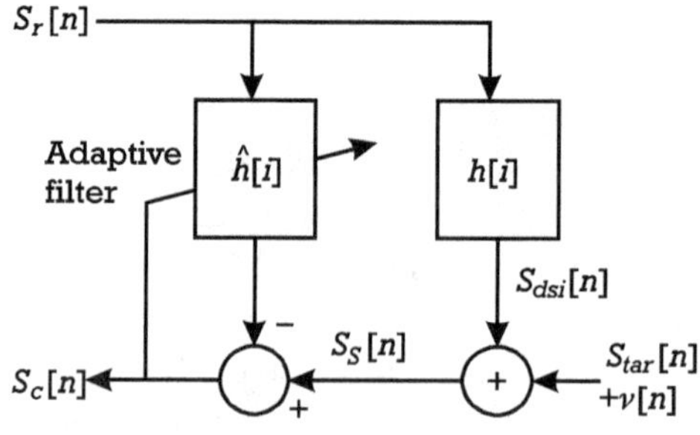

Figure 4.3 Block diagram for direct signal interference suppression.

signal interference-free clean surveillance channel consisting of the target response with additive noise.

While relatively weak clutter responses may have a negligible impact on the total direct signal interference power of (4.4), their power is often significantly greater than thermal noise and consequently will degrade the level of achievable suppression. Therefore, they must be accounted for while estimating $\hat{h}$. The effectiveness of the direct signal interference subtraction method degrades if the number of coefficients is insufficient to properly model the clutter response, but computational requirements increase with channel length. Note that it remains the case that the location at which the direct signal interference is sampled is not coincident with the location of the surveillance antenna phase center and hence there is an unavoidable degradation that is further dependent upon the siting of the receiver system. This is in addition to the complication that arises from multiple directions of arrival of the direct signal due to local multipath.

There has been much research into adaptive signal processing methods for direct signal suppression in passive radar and the required adaptive signal processing is a well-established field encompassing a number of different techniques with many different variants [e.g., 2–6]. Thus, here, the broad method types and their relative merits are considered further.

One cancellation approach designed specifically for passive radar DSI suppression is the Extensive Cancellation Algorithm (ECA) developed by Colone [2]. In this algorithm, data is processed in short time batches and subsequently recombined. This results in a wider cancellation notch in the Doppler domain, thus yielding an improved removal of the direct signal disturbance. This approach is extended over consecutive stages to progressively detect the strongest delay and frequency-shifted replicas of the direct signal and thence reduces their effect on the resulting processed received signal. In this way, the algorithm operates by first removing the contaminating direct signal and the strongest clutter echoes in the surveillance channel. It then detects the strongest peaks in the range-Doppler plane in order of descending signal strength. Appropriate criteria are postulated for the selection of an adequate stopping condition. This leads to an algorithm that is very robust and progressively detects targets, including those with weak echoes that were initially masked by ground clutter and stronger target echo sidelobes. The resulting technique is characterized by a significant improvement in the detection performance, since the joint exploitation of multiple batches and multiple stages allows both a stronger clutter/multipath cancellation (due to the short-time filter weights update) and the capability to extract some of the weak targets echoes that are likely to be lost by conventional, single-stage techniques.

This algorithm can be thought of as a generalized least squares filter, which includes Doppler shifts of the transmitted waveform. However, the method is computationally complex, although there is a simplification via an iterative method that also relies on calculation of the range-Doppler map between each stage. Variants of the CLEAN algorithm originally used with radio astronomy have also been proposed for passive radar processing [3, 4]. These techniques make up block processing techniques that update the subtraction coefficients once per coherent processing interval. Adaptive filters also include the following: normalized least mean squares (NLMS), recursive least squares (RLS), and fast block least squares (FBLS).

A survey comparing Wiener filtering, LMS, and RLS for European Digital Video Broadcasting–Terrestrial (DVB-T) waveforms was performed by Palmer and Searle [5], but the survey does not include the FBLS algorithm, which has recently been proposed but not fully evaluated against other methods of suppression [6]. Absolute target strength was also not discussed and is a parameter that should be included. From recent testing, this algorithm has shown to have excellent performance and run times far faster than the other techniques, lending itself as one of the most promising algorithms for computationally efficient real-time passive radar implementation on stationary platforms. The findings reported in [7] provided a useful comparison of techniques in which the following algorithms are compared:

1. No direct signal interference suppression;
2. Wiener filtering or least squares (WF);
3. CLEAN;
4. Normalized least mean squares (NLMS);
5. Fast block least mean squares (FBLMS);
6. Recursive least squares (RLS).

This comparison provides a good overview and coverage of adaptive and block processing schemes, and allows an informed decision for selecting direct signal interference suppression approaches to be made when designing a passive radar system.

Here, each technique is briefly examined qualitatively and through the resulting range-Doppler maps some insight can be gained into the implied performance and the variation in performance of different approaches. As an example, Figure 4.4 shows a range-Doppler map where no direct signal interference suppression has been applied.

Figure 4.4 shows the full extent of the noise floor across the majority of the image. The noise is sitting at a level of around –60 dB and targets sitting below this will be masked. Indeed, if this is compared to the range-Doppler map shown in Figure 4.5, where a Wiener filter has been applied the noise floor has been pushed down to an average level of approximately –80 dB. Now, not only is the structure of tdhe underlying clutter close to zero Doppler revealed but also a target with a delay time of around 0.02 ms and a Doppler of –0.37 kHz is clearly visible and a solid candidate for automatic detection.

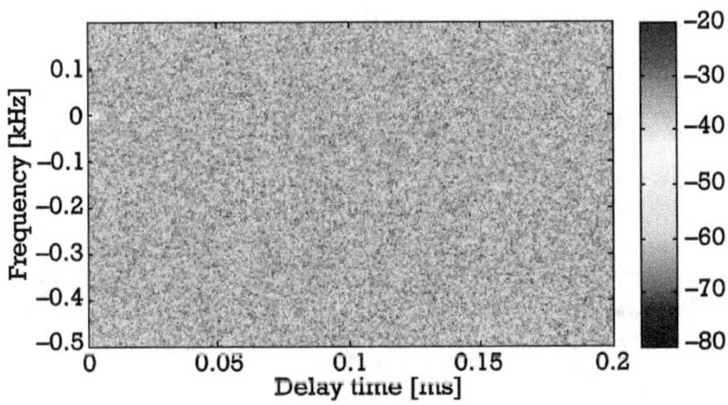

Figure 4.4 A range-Doppler map with no direct signal interference suppression.

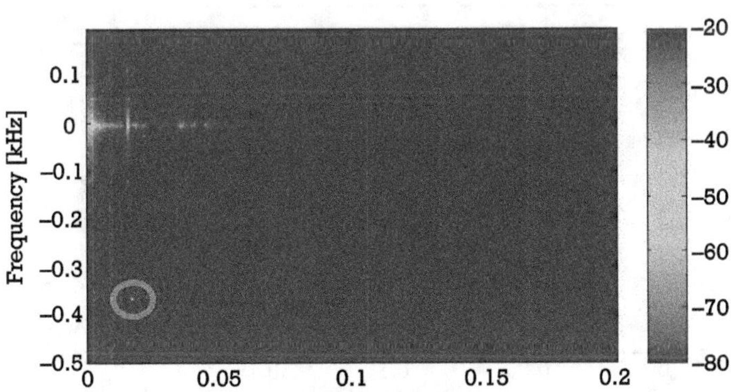

Figure 4.5 A range-Doppler map after Wiener Filter, with target for analysis.

Figures 4.4 and 4.5 also suggest metrics for a quantitative comparisons [7]. These are the maximum direct signal interference level, the noise floor level, the strength of the target echo signal, and the target strength-to-noise floor level (SINR). These values are measured in dB relative to the value at the zero-Doppler, zero-delay location on the range-Doppler map. Collectively these metrics can be used to estimate detection range, especially using measurements taken from operational systems.

Figures 4.6 and 4.7 show a qualitative comparison of the Wiener filter generated range-Doppler map and the FBLMS range-Doppler map concentrating on the area around zero Doppler. Here, the width and depth

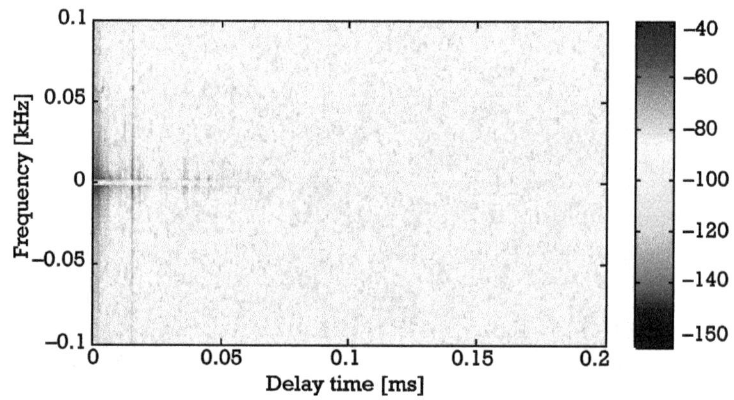

Figure 4.6 The Wiener filter range-Doppler map around zero-Doppler.

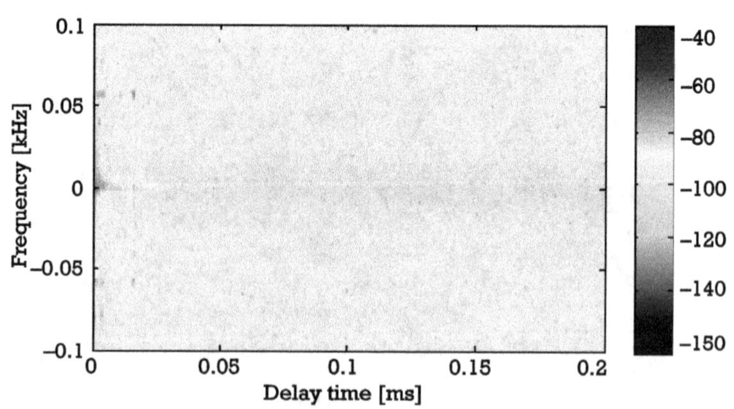

Figure 4.7 The FBLMS filter range-Doppler map around zero-Doppler.

of the nulls formed and their behavior around zero-Doppler have a significant factor in determining overall detection performance of passive radar systems. They also serve to highlight the difficulty in reducing the suppression of the direct signal to a simple metric or set of metrics. The Wiener filter shows a well-controlled suppression of zero Doppler to a level of −110 dB. The background noise floor has a residual value of approximately −85 dB. The behavior of the FBLMS filter is quite different. The filter exhibits a much broader notch in Doppler frequency. This has the consequence that the high zero Doppler suppression is spread significantly to positive and negative Doppler values, possibly resulting in the loss of target detections where radial velocities are low. The FBLMS filter, however, suppresses the background noise below that of the Wiener filter, down to an average level of around −90 dB. This means that there is a better detection performance for target with higher radial velocities. Clearly the effects of different direct signal interference filter schemes are complex and there are trade-offs between the different techniques that have to be carefully considered as part of passive radar design. Table 4.1 [7] provides a quantitative comparison of the main different DSI suppression filters in terms of the four metrics.

One further aspect of direct signal interference filtering that needs to be factored into the choice of scheme is the computational requirements. Although the use of high-speed FPGA processors enables more sophisticated approaches to be used, efficiency and cost will remain design considerations. Digital filtering for improved direct signal interference suppression remains an important area of passive radar development and one that could easily differentiate seemingly similar systems.

Table 4.1
Quantitative Comparison of Direct Signal Interference Suppression

Suppression Scheme	Maximum Direct Signal Interference Level (dB)	Noise Floor (dB)	Target Strength (dB)	SINR (dB)
No Suppression	0.0	−61.3	NA	NA
FBLMS	−49.9	−89.2	−52.3	36.9
Wiener	−82.6	−86.2	−52.5	33.7
CLEAN	−18.2	−70.5	−51.3	19.2
NLMS	−51.4	−89.6	−52.9	36.7

4.4 Summary

In this chapter, it has been seen how passive radar systems must be able to detect the presence of a target many orders of magnitude weaker than the direct signal interference. Due to the continuous nature of most passive radar signals, this source of self-inflicted interference, rather than thermal noise, is likely to determine the sensitivity of the system. Suppression of direct signal interference and clutter prior to range-Doppler processing is crucial for maximizing the effective dynamic range, thus increasing detection range and improving overall system performance. A number of techniques for suppressing the level of directly received signal in the surveillance channel are in existence. In particular, a small number of adaptive filtering techniques have been reported that mitigate the effects of direct signal interference, with varying levels of success. The fast block least mean squares (LMS) filter has been reported as significantly advantageous in terms of suppression performance and has low computational requirements [7]. Practical metrics, such as amount of suppression, run time, and ease of implementation, also serve to help with selection of direct signal interference mitigation algorithms for experimental systems.

References

[1] Monzingo, R. A., R. L. Haupt, and T. W. Miller, *Introduction to Adaptive Arrays*, 2nd ed., Raleigh, NC: SciTech Publishing, 2011.

[2] Colone, F., et al., "A Multistage Processing Algorithm for Disturbance Removal and Target Detection in Passive Bistatic Radar," *IEEE Trans. on Aerospace and Electronic Systems*, Vol. 45, No. 2, April 2009, pp. 698–722.

[3] Feng, B., et al., "An Effective CLEAN Algorithm for Interference Cancellation and Weak Target Detection in Passive Radar," *APSAR 2013*, Tsukuba, Japan, September 23–27, 2013, pp. 160–163.

[4] Kulpa, K., "The CLEAN Type Algorithms for Radar Signal Processing," *2008 Microwaves, Radar and Remote Sensing Symposium*, Kiev, Ukraine, September 22–24, 2008, pp. 152–157.

[5] Palmer, J. E., and S. J. Searle, "Evaluation of Adaptive Filter Algorithms for Clutter Cancellation in Passive Bistatic Radar," *2012 IEEE Radar Conference*, Atlanta, GA, May 7–11, 2012, pp. 0493–0498.

[6] Xiang, M. S., et al., "Block NLMS Cancellation Algorithm and Its Real-Time Implementation for Passive Radar," *IET Int. Radar Conf. 2013*, Xi'an, China, April 14–16, 2013.

[7] Garry, J. L., C. J. Baker, and G. E. Smith, "Direct Signal Suppression for Passive Radar," *ISE Signal Processing Symposium*, Debe, Poland, June 10–12, 2015.

CHAPTER

5

Contents

Passive Radar Performance Prediction

5.1 Introduction

For any radar system, it is important to be able to accurately predict various aspects of system performance and naturally the same is true for passive radar. In this chapter, a simple method for performance prediction via a sensitivity analysis is presented. Examples are given to illustrate the range of possible performance using analog very high frequency (VHF) transmissions, but the methods apply equally well to other forms of illuminators. Other aspects relating to performance prediction are also introduced such as parameter estimation for tracking. However, a detailed review of tracking prediction is beyond the scope of this text. Lastly, published research prediction research where direct comparisons between predicted and measured performance have been made is also briefly examined.

5.2 Detection Performance Prediction Parameters

We begin by recalling that the starting point for a sensitivity analysis of the performance of a passive radar system is the bistatic form of the radar equation as presented in Chapter 4. We repeat the equation here before analyzing the various parameters and their effects of computing performance in a little more detail.

$$\frac{P_r}{P_n} = \frac{P_t G_t}{4\pi R_T^2} \sigma_b \frac{1}{4\pi R_R^2} \frac{G_r \lambda^2}{4\pi} \frac{1}{kT_0 BFL} \quad (5.1)$$

where P_r is the received signal power, P_n is the receiver noise power, P_t is the transmit power, G_t is the transmit antenna gain, R_T is the transmitter-to-target range, σ_b is the target bistatic radar cross-section, RR is the target-to-receiver range, G_r is the receive antenna gain, λ is the signal wavelength, k is Boltzmann's constant, T_0 is the noise reference temperature, 290K, B is the receiver effective bandwidth, F is the receiver effective noise figure, and L are the system losses.

In using this equation to predict the performance of a passive radar system, it is important to understand that each of these parameters should be considered specifically for passive radar design and to understand how appropriate values should be determined when predicting performance. Thus, we examine each of these in turn, considering the range of values that they may take.

5.2.1 Transmit Power

The transmit power P_t can be substantial for many transmission sources that passive radar is able to exploit. For example, broadcast and communications receivers often have inefficient antennas and poor noise figures and the transmission paths are often far from line-of-sight; thus the transmit powers have to be significantly higher to overcome the inefficiencies and losses. Table 3.1 in Chapter 3 gives an overview of some of the common waveforms that can be exploited in passive radar designs. In the United Kingdom, the highest power frequency modulation (FM) radio transmissions are 250 kW (EIRP) per channel, with many more of lower power [1]. The highest-power analog TV transmissions are 1 MW (EIRP) per channel [1]. These are omnidirectional in azimuth and are sited on tall masts on high locations to give good coverage. The vertical-plane radiation patterns are tailored to avoid wasting too much power above the horizontal.

GSM cell phone transmissions in the United Kingdom are in the 900-MHz and 1.8-GHz bands. The modulation format is such that the downlink and uplink bands are each of 25-MHz bandwidth, split into 125 frequency division multiple access (FDMA) channels each of 200-kHz bandwidth, and a given base station will only use a small number of these channels. Each channel carries 8 signals via time division multiple access (TDMA), using Gaussian minimum shift keying (GMSK) modulation. Third generation (3G) transmissions are in the 2-GHz band, using code division multiple access (CDMA) modulation over 5-MHz bandwidth. The radiation patterns of cell phone base station antennas are typically arranged in 120° azimuth sectors, and shaped in the vertical plane again to avoid wasting power. The pattern of frequency reuse means that there will be cells using the same frequencies within very short ranges. Licensed transmit powers are typically in the region of 26 dBWi (26 dB with respect to 1W being transmitted isotropically), although in some cases the actual transmit powers are lower. The Ofcom sitefinder Web site [2] gives details of the location and operating parameters of each base station throughout the United Kingdom and is a useful resource.

In any of the cases that could be considered, it is necessary to have knowledge of the power in the portion of the signal spectrum used for passive radar purposes, which may not be the same as the power of the total signal spectrum. For example, the ambiguity properties of the full signal may not as favorable as those of a portion of the signal. Indeed, this is the case for, say, an analog television transmission, as seen in Chapter 3. The full signal has pronounced ambiguities associated with the 64-μs line repetition rate, but better ambiguity performance may be realized by taking just a portion of the signal spectrum at the expense of reduced signal power.

5.2.2 Target Bistatic Radar Cross-Section

In passive radar, target detection and location are a function of the spatially dependent bistatic radar cross-section, target orientation, target dynamics and the radar design parameters. Targets can be detected in range, Doppler and angle using conventional processing approaches. The target bistatic radar cross section σ_b will not in general be the same as the monostatic cross-section, although for nonstealthy targets the range of values may be comparable [3, 4]. However, rather little has appeared in the literature on the bistatic radar cross section of targets, and this remains an area for future research. In addition, there have only been few reports of bistatic clutter measurements (e.g., [5–7]) and a much more complete treatment is required to enable more realistic calculations of passive radar performance.

Overall, the representation of bistatic radar cross-section for a given target, is not straightforward. Until there are more widely published results, the easiest way to make a choice is to use the bistatic equivalence theorem [8] so that the monostatic equivalent can be used and selected where more widely published data sets are likely to provide an answer.

As the bistatic angle is increased to 180° the region known as forward scatter is encountered. In this region target cross-sections can be considerably enhanced. As seen in Chapter 4, low frequencies are more favorable for the exploitation of forward scatter, so that target detection may be achieved over an adequately wide angular range. This implies that VHF and ultrahigh frequency (UHF) that are often used for passive radar systems are well suited to exploit the forward scatter effect. However, the high transmitter powers mean that the direct signal can swamp the forward scatter component. In addition, the requirement for a baseline crossing is not always met with transmitters that are located on the ground and targets that located in the air. Together these two factors limit the conditions over which forward scatter can be usefully exploited. Kabakchiev et al. [9] demonstrated forward scatter for maritime target detection but used a dedicated transmitter. Forward scatter does not enable range to be measured directly; however, target location can be estimated using a combination of Doppler and bearing as explained in [10].

Another mechanism for enhancement of bistatic radar cross-section of aircraft targets is specular reflection from the underside of the aircraft. This, however, would depend on the specular condition being met and would therefore be ephemeral in nature. This may improve height sensitivity given that most transmitters direct their signals towards the Earth's surface.

5.2.3 Receiver Noise Figure

The noise figures of receivers at VHF and UHF will be of the order of a few decibels at most, so the noise level will be dominated by external noise, most likely in the form of the direct signal, multipath, and other cochannel signals. Unless steps are taken to suppress these signals, the sensitivity and dynamic range of the system will be severely limited.

A simple expression for the amount of direct signal suppression required can be formulated by calculating the ratio of the indirect received signal to the direct signal. Setting this ratio to place the direct signal breakthrough at the same level as receiver noise means that the system can be said to have noise limited, rather the direct signal limited, detection range as is the case for most radar designs. The simple assumption is made that a target can be seen above this level of direct signal breakthrough and hence that it approximates to the highest level of interference that is tolerable for single pulse-like detection as is usual for monostatic radar. However, the benefit from integration may be limited as the direct leakage will also integrate up and as a consequence may lead to a more stringent requirement needing to be set in practice.

Thus, to achieve adequate suppression and hence maintenance of full system dynamic range, the direct signal must be canceled by an amount given by the magnitude of the ratio of the indirect and directly received signals, for example,

$$\frac{P_r}{P_d} = \frac{R_b^2 \sigma_b}{4\pi R_1^2 R_2^2} > \frac{P_r}{P_d} \tag{5.2}$$

where P_r is the target echo signal, P_d is the direct signal, and R_b is the transmitter-to-receiver range (bistatic baseline). This expression is only approximate and strictly speaking the direct signal should be below that of the noise floor after integration, if integration is employed.

In addition, the direct signal can enter the receiving antenna, at very strong levels, through indirect paths. This is due to local reflections from objects such as nearby buildings. One way to deal with this is to use an array-based antenna that forms nulls at all such angles from which these copies of the direct signal arrive.

Consider taking the numerical example of using the real television transmitter, located at Crystal Palace in the south of London, a receiver located at University College London, and assuming a 10 m^2 radar cross-section target and requiring a maximum detection range of 100 km. This equates to a requirement for suppression of direct signal leakage of some 120 dB. It should be noted that as the detection range is reduced from the maximum the amount of direct signal breakthrough compared to the indirect signal will fall. In addition, the leakage signal will be time-varying and subjected to multiple scattering paths. This behavior requires a thorough and detailed understanding to optimize the performance of a given design.

As seen in Chapter 4, there are several techniques that may be used to suppress this leakage. The combination of high-gain antennas and adaptive beamforming also enables multiple simultaneous transmissions to be exploited.

5.2.4 Integration Gain

In a passive radar system, the direct signal is used as a reference against which the indirect or reflected signal can be correlated to provide processing gain for improved sensitivity. Both signal duration and bandwidth contribute to the processing gain in a way that is equivalent to a matched filtering process. The effective receiver bandwidth B (usually the transmission bandwidth) is matched to that of the directly received signal. Thus, this bandwidth combined with a coherent integration time, Tmax, sets the total passive radar matched filter gain at BTmax. A typical bandwidth for a DVB-T signal of 7 MHz coupled with a 1-second coherent integration time leads to an impressive processing gain of 68 dB. Alternatively, a VHF frequency modulation (FM) radio waveform of bandwidth 50 kHz, with an integration time of 1 second, will have a still very large processing gain of 47 dB. Indeed, it is processing gains of this magnitude that enable such passive radar systems to have substantial detection ranges of over 400 km (target to receiver range) when using powerful VHF transmitters.

However, this processing gain is limited, through restrictions on the coherent integration time through two main effects, range and velocity cell migration. Both forms of migration are a maximum when a target moves radially towards the transmitter or receiver. Bistatic acceleration can also cause velocity and range migration. An approximate rule of thumb for the maximum value of the coherent processing gain is

$$T_{MAX} = \left(\frac{\lambda}{A_R} \right)^{1/2} \tag{5.3}$$

where A_R is the radial component of target acceleration. This expression is derived from the Newtonian equation of motion $s = ut + ½\, at^2$. The $½\, at^2$ term represents the distance traveled due to an acceleration A over a time t, and setting this to be equal to $\lambda/2$, corresponding to a change in phase of 360°, gives (5.3). Some authors have used a version of this equation with a factor $\sqrt{2}$ in the denominator, reducing the coherent integration time by approximately 40% [11] but ensuring a high degree of gain. Either approximation is valid and reflects uncertainty in what may occur in practice. With either assumption, the maximum processing gain may be written as

$$G_p = T_{MAX} B \tag{5.4}$$

The maximum processing gain also depends on the time for which the target echoes remain coherent. The complex nature of most man-made targets, such as aircraft, means that over time, as they present orientation changes, the scattering received at the radar becomes increasingly incoherent. Typical processing times for FM and DVB-T passive radar systems lie in the region of 0.1 to 1 second as a compromise between range and velocity cell migrations and target coherence.

5.2.5 System Losses

Passive radar losses are no different from any other radar and result from system induced losses or due to a variety of propagation effects. One difference in which passive radar losses have to be treated with care is caused by the particular illumination patterns of transmission sources. For example, FM or DVB-T illuminators are designed to cover a region on the ground and as a direct consequence minimize radiation in the vertical plane. However, antenna designs, in combination with the relatively long wavelengths used, means that there is substantial radiation emitted above the ground. It is this that is used for aircraft detection, but at the same time it means that power densities cannot be considered to be omnidirectional and overall detection ranges will be less than if an omnidirectional assumption is made. In many cases, beam patterns are available from national agencies and can be used to gain a better estimate of radar range. This is also true of other illuminators, and Gao et al. [12] allowed for the antenna gain patterns of WiMAX illuminators and their receiver in range computations. The waveform emitted by an illuminator of interest may also be suboptimal for use in passive radar and as a result signal processing losses can occur. This has to be taken into account on a case by case basis for different illumination types. Propagation losses also have to be considered a little differently for passive radar as the siting of the illuminator (and to a lesser extent the receiver) is outside the control of the radar designer. This has a very significant bearing on coverage as illuminators are often ground-based and sited to maximize reception in areas of highest population density. This may not coincide with areas of greatest importance for, say, an FM or DVB-T passive radar being used for air traffic management applications. Propagation models exist for many illuminators of opportunity and are also available from national agencies. However, they can be subject to significant variation. For example, in [13], Barrot compared two models, the Advanced Refractive Effects Prediction System model (AREPS) and the Irregular Terrain Model (ITM). The AREPS and ITM are categorized as point-to-point and point-to-area models. Although they have been developed for similar purposes, their computational approaches are quite different. Further, the results presented show substantial difference in the propagation predictions and hence show significant difference in radar coverage and detection ranges. As a consequence, it has been concluded that great care should be exercised in using such models to ensure that they really are adding credible realism to radar performance prediction.

5.3 Detection Performance Prediction

All of the foregoing has shown that some care must be taken when selecting values for the parameters to be used in the bistatic form of the radar equation to predict the performance of passive radar designs. In this section we present performance predictions for three "straw-man" systems, attempting to show the likely achievable performance and to identify critical factors. The systems considered are FM radio, cell phone base stations, and digital radio. In each case an omnidirectional receive antenna, a noise figure of 5 dB, losses of 5 dB, and full suppression of direct signal leakage are assumed.

FM radio transmissions have the inherent attractive properties of very broad coverage and relatively high transmitter powers. For the example considered here, the BBC transmitter at Wrotham in the southeast of England is taken together with a receiver sited at the Engineering building of UCL in central London. The transmitted power is 250 kW and broadcasts are made in the frequency range of 89.1 to 93.5 MHz. Figure 5.1 shows a plot of the detection range assuming a target with a radar cross-section of 100 m^2, an integration time of 1 second and a modulation bandwidth of 55 kHz. The commencement of the white region represents

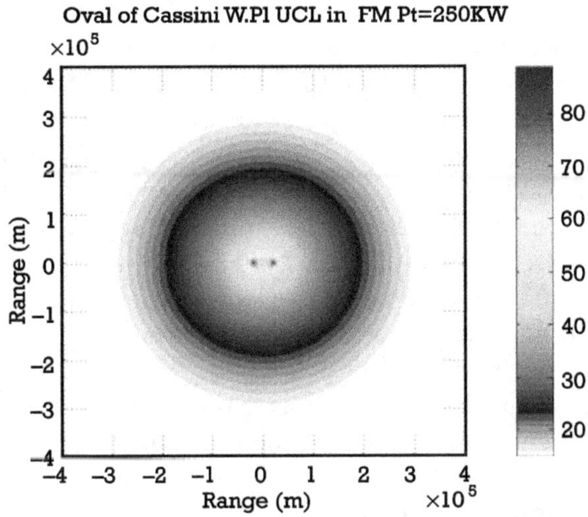

Figure 5.1 Detection range for a transmitter at Wrotham in southeast England and a receiver at UCL.

a contour with a signal to noise ratio of 15 dB (and this is used for all subsequent figures of this type).

Note that the modulation bandwidth is considerably less than that specified for the transmissions. Recalling from Chapter 3 that the modulation bandwidth is a function of media content and is therefore also a function of time, 55 kHz represents a typical overall value for the bandwidth of the transmitted signal to be exploited. A signal-to-noise ratio of 15 dB or greater is maintained out to a range of nearly 300 km. However, this is a free-space calculation and no effects of terrain and propagation have been included. Thus, it is likely to represent a best case and actual performance would be expected to be to a reduced detection range. It should also be noted, for the case being considered, that as the maximum detection range is approached, the bistatic system starts to approximate to that of a monostatic system. The Ovals of Cassini approximate to circles and the hyperbolae representing contours of contours of constant Doppler (isodops) approximate radial lines as for monostatic radar. If such an approximation can be legitimately made, it will simplify much of the resulting processing and make bistatic passive radar easier to design and assess. For example, under these conditions it is probably legitimate to use monostatic values for target and clutter signatures. It should be further noted that the power emitted by transmitters across the United Kingdom varies from as little as 4W to a maximum of 250 kW and this variation has to be carefully factored in to performance predictions.

Figure 5.2 shows how the detection range alters when a second, different, transmitter is exploited. Here the transmitter located at Crystal Palace in south London is used, which has a reduced transmit power of 4 kW. As might be expected, the detection range significantly reduced as there is approximately 18 dB less transmitted power. The signal to noise ratio of 15 dB occurs for a detection range of just over 100 km. Figure 5.3 illustrates how the coverage changes when the two transmitters are exploited together using noncoherent integration. Now the combined detection range is extended to over 300 km.

An alternative approach is to process the detections from each transmitter independently and then combine them, as this is simpler. It might be thought, at first sight, that coherent combination would yield the highest integration efficiencies. However, full coherent combination is not usually possible as the transmissions will probably be at differing frequencies and not phase coherent with each other. Overall the high transmit powers and good coverage make FM radio transmissions particularly well suited to air target detection for both civil and military applications. Equally, they could be used for marine navigation in coastal waters although clutter may be a more significant factor.

In a second case, a cell phone base station transmitter is used and the parameters are listed in Table 5.1.

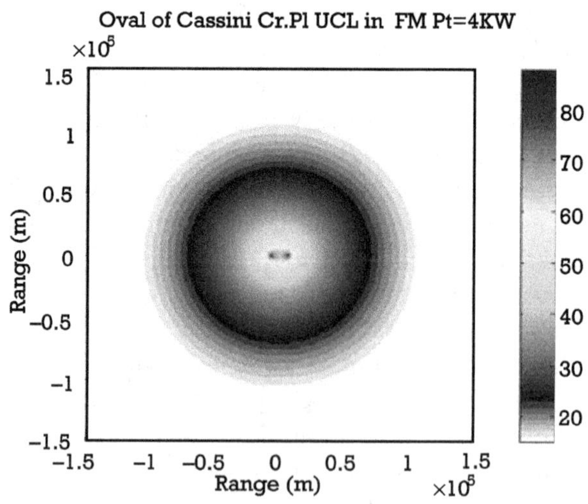

Figure 5.2 Detection range with transmitter at Crystal Palace and the receiver at UCL.

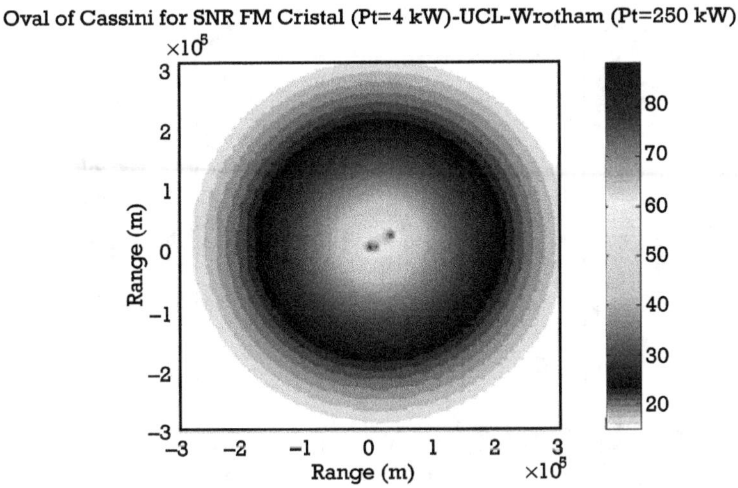

Figure 5.3 Detection range for transmitters at Wrotham and Crystal Palace and a receiver at UCL.

Table 5.1
Example Attributes of a Cell Phone Base Station Located to the Northern End of Gower Street in London, United Kingdom

Name of Operator	T-MOBILE
Operator Site Reference	98463
Height of Antenna	35.8m
Frequency Range	1800 MHz
Transmitter Power	26 dBW
Maximum licensed power	32 dBW
Type of Transmission	GSM

This particular transmitter has an operating frequency of 1,800 MHz and is located towards the northern end of Gower Street approximately 200m from the engineering building of UCL where the receiver is again placed. The other parameters are maintained constant as with the first case. A plot of the detection range is shown in Figure 5.4, which suggests a maximum range of around 12 km.

As might be expected, with a much reduced transmitter power, the predicted detection range is much less than for the previous example. It would seem therefore that such transmitters have much more limited

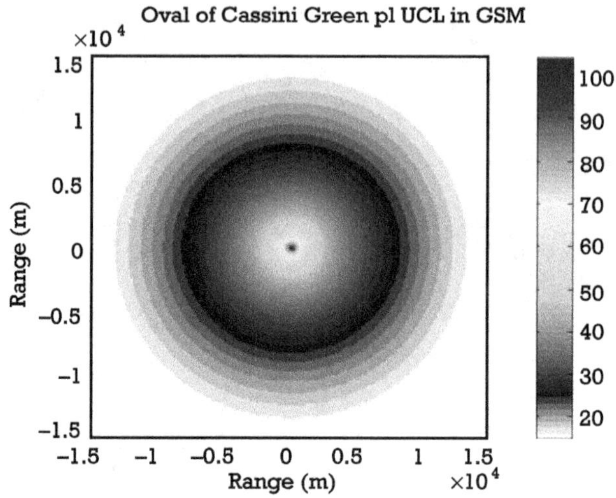

Figure 5.4 Detection range for a cell phone base station located at the northern end of Gower Street in London and the receiver at UCL.

applications. However, as there is such an extensive and diverse network of base station transmitters targets could be tracked though such a network and hence the coverage may be extended greatly. This characteristic could greatly extend the range of applications to include examples such as counting of vehicles for traffic flow management, remote monitoring of movement around buildings as a security device and possibly acting as a cue for a camera system.

The third example uses a digital audio broadcast (DAB) transmission from Crystal Palace. This has a transmit power of 10 kW. Figure 5.5 shows the resulting detection range. As might be expected for a higher-power transmitter, the coverage is extended to a range of around 90 km. However, it should be noted that despite the transmit power being higher than for the FM transmission at Crystal Palace, the maximum detection range is shorter. This is due to the higher frequency offsetting the lower transmit power. Again, it should be noted that output powers of transmissions of this kind vary between 500W and 10 kW. Additionally, coverage from a single transmitter site is not currently as universal as is the case for FM transmissions, although new transmitters are constantly being added. It can also be seen that the ways in which multiple transmitters and frequencies can be combined will have a performance very much dictated by the particular parameters of the transmitters selected.

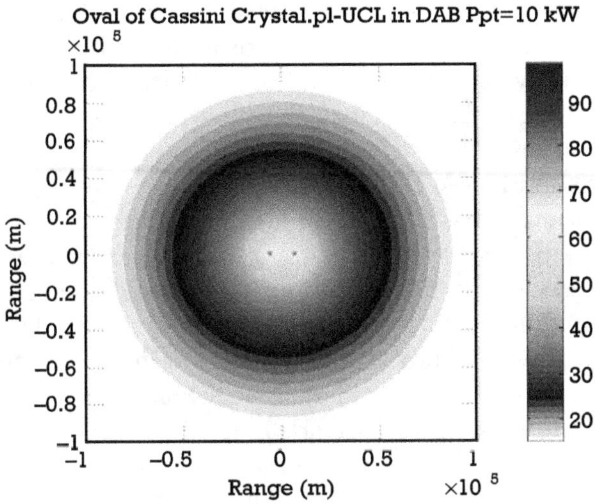

Figure 5.5 Detection range for a DAB transmitter at Crystal Palace and a receiver at UCL.

As already noted, one of the strengths of passive radar design is that several different types of emission of opportunity could be exploited at a single receiver site. This has the advantage of providing frequency diversity and spatial diversity and as such makes passive radar somewhat equivalent to a multisite or netted radar system (i.e., multiple transmitter locations and a single receiver).

Overall, it is possible to predict the performance of a passive radar design with a moderately high degree of accuracy and confidence provided due care has been exercised in selecting the parameters used in the computation. Nonetheless, the results of such predictions should be taken as indicative of expected performance and such computations allow the effects of major design choices such as the effects of different receiver site locations to be assessed. In the next section, some examples are examined where a comparison has been made between predicted and experimentally measured performance.

5.4 Comparing Predicted and Experimental Detection Performance

Perhaps the most comprehensive investigation of detection performance in which model predictions are compared with experimentally evaluated performance has been carried out by Malanowski et al. [14]. In this paper, they used the radar equation as constructed for passive radar and consider the specific case of using an FM transmitter of the type used in their experiments. They also take the illumination coverage into account and compute dynamic range as well as detection range. The maximum free-space detection range is calculated to be 440 km and a discussion of the various effects considered in the previous section follow. The experiments use the PaRaDe FM-based passive radar (see Chapter 7) and a bistatic range of 700 km was achieved resulting in a target to receiver range of approximately 350 km. It is suggested that the difference between the free space computed detection range and the measured detection range is due to a number of factors. These include interference in the FM band, uncertainty in the true target radar cross-section, propagation, and multipath losses. However, the receiver range of 350 km for an FM-based passive radar using a single transmitter as an illuminator of opportunity is astonishingly good and provides a solid indicator of the potential for passive radar in air traffic and air defense applications.

5.5 Target Location

The prediction of passive radar tracking performance has received less attention than its detection counterpart but is nevertheless of central importance in many applications. Traditional monostatic radars will use a combination of range, angle (elevation and azimuth) and velocity resolution to provide a good indication of target location and velocity followed by improved location estimation using techniques such as early-late gate and monopulse. These estimates contain measurement noise, which is smoothed using, say, a Kalman filter with appropriately adjusted parameters. FM-based passive radar has some additional characteristics that affect tracking performance. FM-based passive radar has relatively poor angular and range resolutions but very good Doppler resolution (due to the long coherent integration times). If targets can be resolved such that only one occurs in each of these four-dimensional (4-D) resolution cells, then they will be unambiguous and a more refined location estimation can take place. Malanowski and Kulpa [15] compared a number of estimation techniques where only a single target is considered and present tracking accuracy results. They show how tracking accuracies are dependent on target integration time, which, in turn, determines measurement noise through its effect on the signal-to-noise ratio.

5.6 Advanced Passive Radar Performance Prediction

One of the inherent features of many of the possible passive radar illuminators is that there are often multiple transmitters available and each may have multiple frequencies of transmission. In [16–18], performances for experimental systems that exploit multiple frequencies are reported and in [19] the performance under MIMO operation is computed, all using an FM-based passive radar configuration. In particular, Han and Inggs [19] showed, using a Neyman-Pearson hypothesis, that a closed form expression for the probability of detection can be derived. They use this to demonstrate that the detection performance improves as a function of the number of transmitters and receivers in a multiple input, multiple output (MIMO) passive radar. The computation is somewhat idealized, but nevertheless together with the experimental results reported in [16–18] indicates that there is considerable further scope for enhancing performance in passive radar and overcoming some of the losses and drawbacks that result from using illuminators designed for other purposes.

Lastly, it was noted that in [20] Tan et al. reported an analysis of target detection performance for an airborne passive radar concept. They highlighted that performance is highly dependent on geometry, which is continuously changing for an airborne receiver. They also showed that the bistatic ground clutter power is considerably reduced (as the ground is further away) and hence suppression of the directly received signal is critical if such a system is to have operationally useful detection ranges. It should be noted that in [21] Brown reported on experimental FM airborne passive radar results and highlighted the importance of the direct signal term while demonstrating that airborne operation is a reality.

5.7 Summary

In this chapter, we have recast the bistatic radar equation into a form that readily reflects the design features of a passive radar system. This highlights the importance of the bistatic geometry and the key dependence on the nature of the illuminating waveform. The form and nature of bistatic reflections are not well known and require extensive further research. A rule-of-thumb expression has been presented that indicates the high levels of direct signal suppression that are required to ensure that maximum detection ranges can be achieved.

Prediction of detection range and coverage for a variety of illuminators of opportunity shows that receiver detection ranges of up to 350 km are possible. This is, again, highly dependent on the properties of the illuminator. However, it is expected that full-scale systems will have a performance near to the levels predicted here. Thus, passive radar may be invoked to support quite a wide range of applications provided they are consistent with the limitations imposed by the availability of illuminators. Indeed, the plethora of radio frequency radiation sources will undoubtedly increase still further and the case for passive radar becomes ever more compelling. Furthermore, sophisticated processing techniques such as SAR, ISAR, interferometry, and others can all be exploited, and examples of these are described in Chapter 7. This allows ambiguity function properties to be evaluated and examined. In this way, further realism can be applied to predictions of actual radar performance and consequently a sound foundation for a much more comprehensive approach to passive radar system design can be established.

References

[1] http://www.bbc.co.uk/reception/, accessed October 17, 2016.

[2] http://www.sitefinder.ofcom.org.uk/, accessed October 17, 2016.

[3] Jackson, M. C., "The Geometry of Bistatic Radar Systems," *IEE Proc.*, Vol. 133, Pt. F, No. 7, December 1986, pp. 604–612.

[4] Kell, R. E., "On the Derivation of Bistatic RCS from Monostatic Measurements," *Proc. IEEE*, Vol. 53, August 1965, pp. 983–988.

[5] Larson, R. W., et al., "Bistatic Clutter Measurements," *IEEE Trans. on Antennas and Propagation*, Vol. AP-26, No. 6, 1978, pp. 801–804.

[6] Wicks, M., F. Stremler, and S. Anthony, "Airborne Ground Clutter Measurement System Design Considerations," *IEEE AES Magazine*, October 1988, pp. 27–31.

[7] McLaughlin, D. M., et al., "Low Grazing Angle Bistatic NRCS of Forested Clutter," *Electronics Letters*, Vol. 30, No. 18, September 1994, pp. 1532–1533.

[8] Willis, N. J., *Bistatic Radar*, Raleigh, NC: SciTech Publishing, 2005.

[9] Kabakchiev, C., et al., "CFAR Detection and Parameter Estimation of Moving Marine Targets Using Forward Scatter Radar," *12th International Radar Symposium*, Warsaw, September 2011, pp. 85–90.

[10] Howland, P. E., "Target Tracking Using Television Based Bistatic Radar," *IEE Proc Radar, Sonar and Navigation*, Vol. 146, No. 3, 1999, pp. 166–174.

[11] Malanowski, M., and K. Kulpa, "Analysis of Integration Gain in Passive Radar," *2008 Int. Radar Conference*, Adelaide, Australia, September 2–5, 2008, pp. 323–328.

[12] Gao, G., Q. Wang, and C. Hou, "Power Budget and Performance Prediction for WiMAX Based Passive Radar," *6th Intl. Conference on Pervasive Computing and Applications*, Port Elizabeth, South Africa, October 26–28, 2011, pp. 517–520.

[13] Dabrowski, T., W. Barott, and B. Himed, "Effect of Propagation Model Fidelity on Passive Radar Performance Predictions," *2015 IEEE Int. Radar Conference*, Arlington VA, May 10–15, 2015, pp. 1503–1508.

[14] Malanowski, M., et al., "Analysis of Detection Range of FM-Based Passive Radar," IET Radar, Sonar and Navigation, Vol. 8, No. 2, 2014, pp. 153–159.

[15] Malanowski, M., and K. Kulpa, "Analysis of Bistatic Tracking Accuracy in Passive Radar," 2009 IEEE Radar Conference, Pasadena CA, May 4–8, 2009.

[16] Malanowski, M., et al., "Experimental Results of the PaRaDe Passive Radar Field Trials," *13th International Radar Symposium*, May 23–25, 2012, pp. 65–68.

[17] Edrich, M., and A. Schroeder, "Design, Implementation and Test of a Multiband Multistatic Passive Radar System for Operational Use in Airspace Surveillance," *2014 IEEE Radar Conference*, Cincinnati, OH, May 19–23, 2014, pp. 0012–0016.

[18] Bongionni, C., F. Colone, and P. Lombardo, "Performance Analysis of a Multi-Frequency FM Based Passive Bistatic Radar," *2008 IEEE Radar Conference*, Rome, Italy, May 26–30, 2008.

[19] Han, J., and M. Inggs, "Detection Performance of MIMO Passive Radar Systems Based on FM Signals," *CIE International Conference on Radar*, Vol. 1, Chengdu, China, October 24–27, 2011, pp. 161–164.

[20] Tan, D. K. P., et al., "Target Detection Performance Analysis for Airborne Passive Radar Bistatic Radar," *2010 IEEE International Geoscience and Remote Sensing Symposium*, Honolulu, HI, July 25–30, 2010, pp. 3553–3556.

[21] Brown, J., "FM Airborne Passive Radar," Ph.D. thesis, University College London, 2013.

CHAPTER

6

Contents

Detection and Tracking

6.1 Introduction

Many radar applications require targets to be both detected and tracked. Detection constitutes the first stage and its objective is to simply declare the presence (or absence) of a target. The next stage is to estimate the locations of detected targets. These locations are then filtered to give an improved and smoothed estimate of target locations and to do this on a continuous basis so that target trajectory histories can be obtained. In other words, detection of targets is not the whole story. In practice the required output from the radar will be tracks of individual targets, showing their evolution and direction. Conventionally this is achieved either using radars dedicated to tracking individual targets, or by track-while-scan processing of detection of multiple targets by a radar which scans in azimuth. The task is then to associate the detections with individual targets and to determine their evolution as a function of time. The situation with passive bistatic radar is a little different, in that the transmit sources are in general omnidirectional. In addition, for the more commonly used very high frequency (VHF) and ultrahigh frequency (UHF) transmissions, the receive antenna beams are typically quite broad, of the order of 90°. This means that the location of targets can be very imprecise and in the absence of angular information can be ambiguous. Also range resolution is comparatively coarse, being of the order of kilometers for VHF and tens of meters for UHF. However, an advantage of the passive radar approach is that track updates can be generated at much more rapid time intervals than for conventional scanning radar. With scanning radar the updates are fixed by the beam revisit time. In passive radar the target is continuously illuminated and echoes are continuously received. This means that there is an almost unlimited choice of update times. The purpose of this chapter is to show how detections in a passive radar may be processed to provide usable target tracks.

6.2 CFAR Detection

Detection approaches in passive radar are able to use conventional approaches and hence we will concentrate more attention on the tracking stage of processing. Nevertheless, to set the scene, a brief review is provided below of one of the more common processing techniques used to detect targets as applied to passive radar. For the interested reader, the introductory textbook, *Principles of Modern Radar,* provides an excellent overview of CFAR detection [1].

The first stage in the processing is detection the presence or absence of a target. One commonly used approach to target detection is to apply a constant false alarm rate (CFAR) detector, setting an adaptive detection threshold based on the surrounding noise and clutter. Howland, in describing a passive radar system based on an FM radio transmitter [2], reports that a conventional cell-averaging CFAR with guard cells either side of the test cell is suitable, with a CFAR window of 10 cells in total. Given the coarse range resolution of the FM radio-based passive radar (B = 50 kHz gives $c/2\,B = 3$ km), this is quite appropriate.

Figure 6.1 shows the raw (left) and CFAR-detected (right) range-Doppler plots for a PBR system based on a single FM radio transmitter and a single receiver, detecting aircraft targets in the vicinity of Pretoria, South Africa. In this way, targets are allocated particular combinations of range and Doppler and these values can be compared with those of the range of possible target locations and many, if not all, of the incorrect values can be

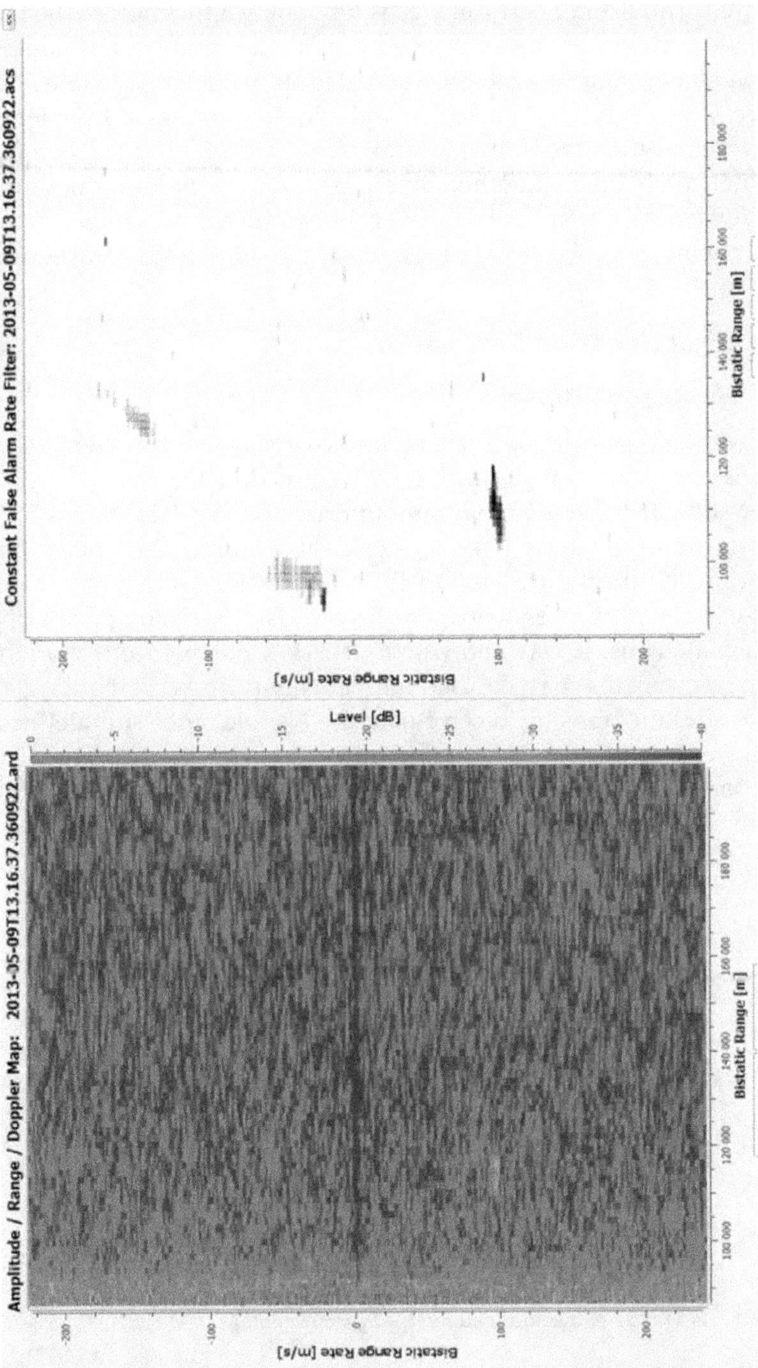

Figure 6.1 Raw (left) and CFAR-detected (right) range-Doppler plots. (Courtesy of Dr. Craig Tong, UCT.)

eliminated. The overall performance is ultimately a function of the transmitter and receiver parameters together with the numbers of targets in the surveillance volumes and their actual locations relative to one another. Once targets have been detected and allocated to range, Doppler, and azimuth cells, estimates of their locations can be made that are then fed to track filter to obtain a smoothed, improved estimate of location. This is a continuous process that builds up over time to reveal the target track history. In the next section, location estimation techniques are introduced.

6.3 Target Location Estimation

6.3.1 Iso-Range Ellipses

The target echo information available at the passive radar receiver for each target may be bistatic range sum $(R_T + R_R)$, direction of arrival (DOA), and Doppler shift. The bistatic range sum information typically comes from a measurement of the delay $(R_T + R_R - L)/c$ between reception of the direct signal from the transmitter and of the target echo, usually by means of a cross-correlation process between the echo and a clean version of the direct signal. Thus, if L is known, $(R_T + R_R)$ is obtained directly. These three measurements may be used individually or in combination. For a particular bistatic transmit/receive pair, the bistatic range sum defines an iso-range ellipse on which the target must lie, with the transmitter and receiver as the two focal points (Figure 6.2). If a measurement of DOA of the echo at the receiver is available, then this unambiguously defines the target location on the ellipse.

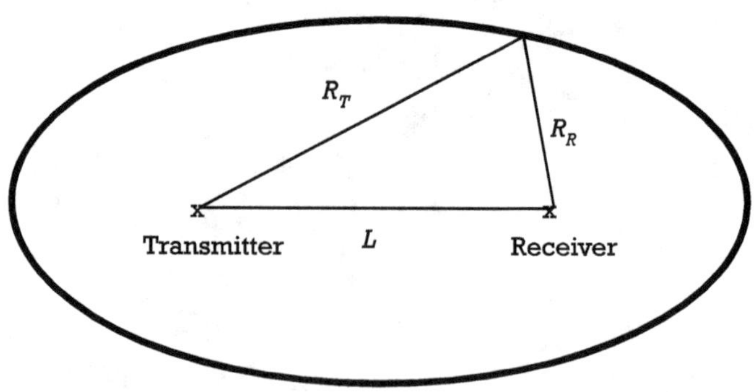

Figure 6.2 Iso-range ellipse defined by $R_T + R_R$ = constant.

Otherwise, if bistatic range information from more than one transmit/receive pair is available, the target location can be determined from the points at which the iso-range ellipses intersect (Figure 6.3).

However, this information is almost always ambiguous because there are several such intersection points, so some means of identifying the correct one and rejecting the others is required. This might take the form of multilateration from further iso-range ellipses (Figure 6.4) or use of Doppler information.

These ambiguous target positions have been called Caspers' Ghosts, after Jim Caspers, who was the author of the chapter on bistatic radar in Skolnik's first *Radar Handbook* [3]. In the general case, for N transmit-receive pairs and n targets, the number of potential ghosts is:

$$\frac{\left(2n^2 - n\right)\left(N^2 - N\right)}{2} \tag{6.1}$$

Perhaps inevitably, the process of identifying and excising ghosts has become known as ghostbusting.

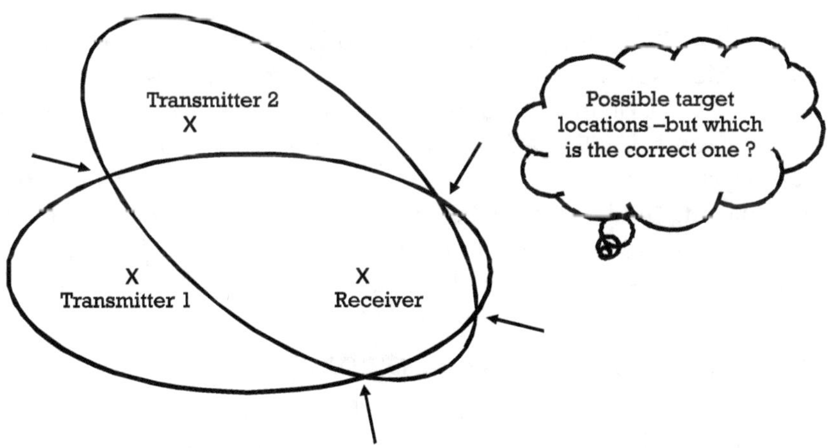

Figure 6.3 Iso-range ellipses corresponding to the bistatic range sums from a single target, for a single receiver and two transmitters. A similar situation would occur for one transmitter and two receivers. The ellipses intersect at four locations of which only one corresponds to the true target location. The others are ghosts.

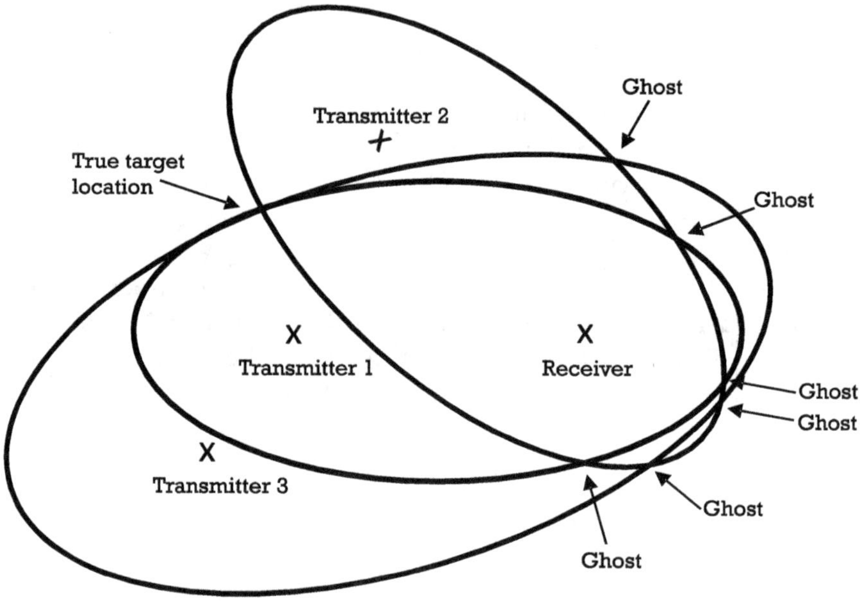

Figure 6.4 Iso-range ellipses corresponding to the bistatic range sums from a single target for a single receiver and three transmitters. The three ellipses intersect at one point, which is the true target location, but in this case there are three other ghost locations where two ellipses intersect.

The use of Doppler information to resolve ghosts is illustrated in Figure 6.5, which is taken from the airborne passive radar system described in Section 7.5 in Chapter 7. Here the receiver is carried by an aircraft, flying over southeast England. Iso-range ellipses corresponding to a particular target and two frequency modulation (FM) radio transmitters at Wrotham and Guildford are labeled A and B. Also shown are Doppler vectors corresponding to the measured Doppler shift, taking into account the known speed and direction of the aircraft carrying the receiver.

6.3.2 Time Difference of Arrival (TDOA)

Another way of approaching target localization is in terms of the time difference of arrival (TDOA) of echoes at pairs of receivers. This is similar in form to a number of problems in radiolocation and navigation. Figure 6.6 shows a situation with a single transmitter and two receivers (by reciprocity, transmitters and receivers are interchangeable). The TDOA of the echoes at the two receivers is just:

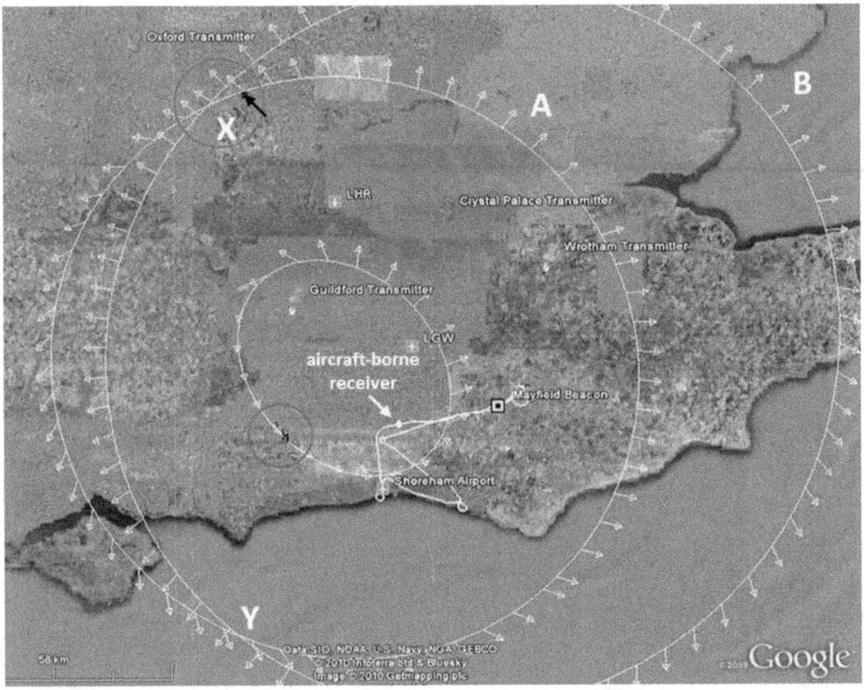

Figure 6.5 Ambiguity resolution by Doppler. Two iso-range ellipses corresponding to a commercial aircraft target and two FM radio transmitters are labeled A and B. The ellipses intersect at two points X and Y. The Doppler vectors at X agree, while those at Y do not, showing that X is the correct target location and Y is a ghost. The black arrow shows the target location and vector obtained from ADS/Mode-S information, providing independent confirmation. (Courtesy of Dr. James Brown [4].)

$$\frac{\left(R_{R_1} - R_{R_2}\right)}{c} \tag{6.2}$$

The contours of constant TDOA are hyperboloids. By combining measurements from several pairs of receivers, the target location can be determined as the point where the hyperboloids intersect. In the general three-dimensional case, a minimum of three receiver pairs is necessary.

Malanowski and Kulpa [5] point out that solution of the problem can be challenging because of the nonlinear relation between the target location and the measured parameters. They derive and analyze two closed-form solutions, which they denote spherical interpolation (SI) and spherical intersection (SX), and evaluate them both by means of simulations

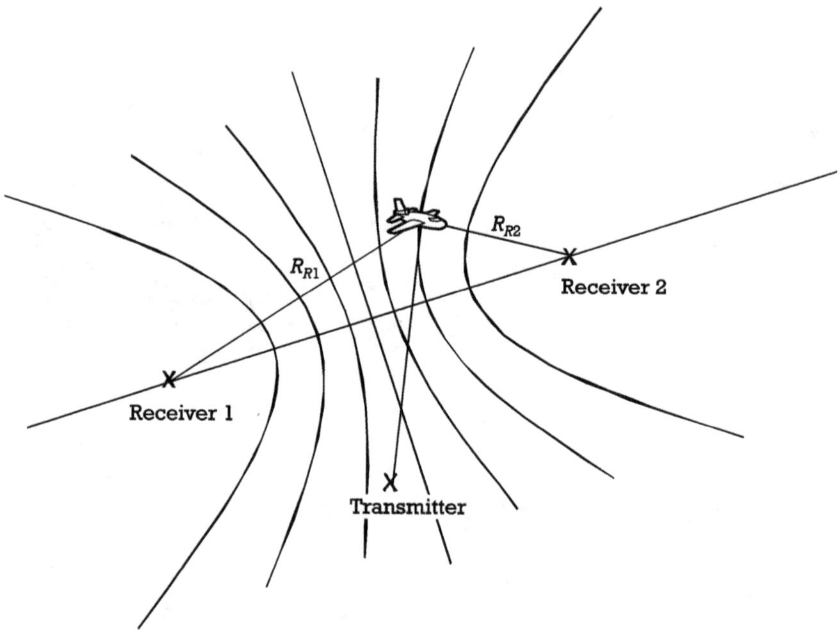

Figure 6.6 The TDOA of the echoes at the two receivers is $(R_{R1} - R_{R2})/c$.

and with real measured data. As an example, Figure 6.7 shows the simulated and theoretical errors for a passive radar consisting of three transmitters (marked by triangles) and one receiver (marked by a circle), as a function of positional error in the x direction. The reader is referred to [5] for full details of the algorithms and results.

6.3.3 Range-Doppler Plots

Separating targets by resolving their combined range and Doppler values greatly helps to remove ambiguity. Passive radar systems typically utilize very long integration times of the order of 1 second. A 1-second integration time is equivalent to a 1-Hz Doppler resolution. This represents a very fine degree of resolution and it is Doppler that takes a key role in removing ambiguities and in providing at least one tracking parameter, target velocity, with high accuracy. Standard techniques such as velocity gating, which employ high- and low-velocity filters, can be applied to improve the estimate of target velocity still further. A common means of presenting the information from the receiver is in the form of a range/

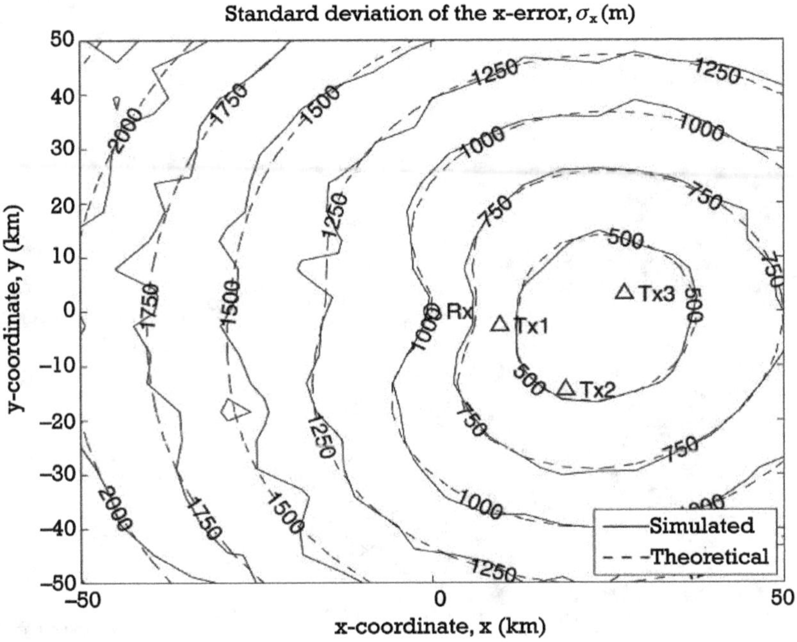

Figure 6.7 Simulated (solid line) and theoretical (dashed line) standard deviation of position error in the x direction [5].

Doppler plot. The Doppler information is obtained by fast Fourier transform (FFT) processing at each range resolution cell over a suitable integration interval, and plotted as a function of range and Doppler. Any residual direct signal and clutter after the suppression techniques described in Chapter 4 show up at zero Doppler and close range, and targets will be visible at the appropriate bistatic range and Doppler. An example of this is shown in Figure 6.8.

6.4 Track Filtering

Having obtained estimates of the location of detected targets (measurement estimates) the next stage is to use a tracking filter to refine those estimated and provide a continuous set of outputs from which target track histories can be observed. Various approaches to tracking filters for passive radar have been described in the literature. All involve the same basic operations of track initiation, association, track confirmation, target state estimation, and track deletion as in conventional radar tracking, and the

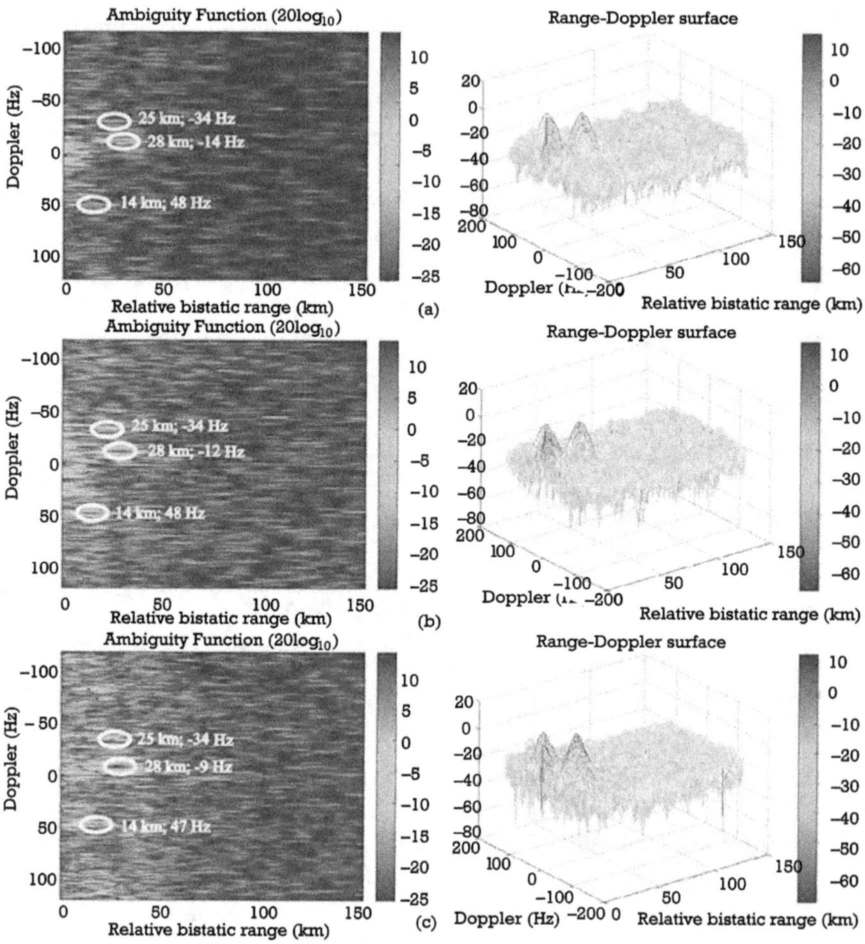

Figure 6.8 Two-dimensional bistatic range-Doppler display plots for the BBC 91.3 MHz Wrotham FM transmitter. (a–c) show three successive 1-second blocks of data [6].

reader is referred to the well-known books on standard and advanced radar tracking techniques [7–10]. Indeed, taking into account the particular characteristics of passive radar outlined above it is possible to adapt conventional tracking approaches. Here, we use an example based upon the published research of Howland [2], which employs, perhaps, the most common form of track filter, the Kalman filter.

6.4.1 Kalman Filter

Howland's approach [2] makes use of both range-Doppler and DOA information in a basic Kalman filter arrangement, based on that described in Section 1.5 of [8]. Using the same description and notation as [2], the measurement vector $z(k)$ is made up of measurements of range R_k, Doppler F_k and bearing Φ_k:

$$z(k) = \left(R_k F_k \Phi_k\right)' \tag{6.3}$$

and the state vector $x(k)$ consists of range, range-rate, Doppler, Doppler-rate, bearing, and bearing rate:

$$\mathbf{x}(k) = \left(r(k)\dot{r}(k)f(k)\dot{f}(k)\phi(k)\dot{\phi}(k)\right)' \tag{6.4}$$

The state transition matrix is:

$$\mathbf{F}(k) = \begin{pmatrix} 1 & 0 & -\lambda\tau & 0 & 0 & 0 \\ 0 & 0 & -\lambda & -\lambda\tau & 0 & 0 \\ 0 & 0 & 1 & \tau & 0 & 0 \\ 0 & 0 & 0 & 1 & 0 & 0 \\ 0 & 0 & 0 & 0 & 1 & \tau \\ 0 & 0 & 0 & 0 & 0 & 1 \end{pmatrix} \tag{6.5}$$

so

$$\mathbf{x}(k+1|k) = \mathbf{F}(k)\hat{x}(k|k) \tag{6.6}$$

where τ is the update interval and λ is the wavelength.

The state prediction covariance matrix is updated according to:

$$\mathbf{P}(k+1|k) = \mathbf{F}(k)\mathbf{P}(k|k)\mathbf{F}(k)' + \mathbf{Q}(k) \tag{6.7}$$

where the state transition matrix is defined in the standard way:

$$F(k) = \begin{pmatrix} 1 & \tau & 0 & 0 & 0 & 0 \\ 0 & 1 & 0 & 0 & 0 & 0 \\ 0 & 0 & 1 & \tau & 0 & 0 \\ 0 & 0 & 0 & 1 & 0 & 0 \\ 0 & 0 & 0 & 0 & 1 & \tau \\ 0 & 0 & 0 & 0 & 0 & 1 \end{pmatrix} \tag{6.8}$$

In this implementation, the association gate is defined by:

$$\left[z - \hat{z}(x+1|k)\right]' \boldsymbol{S}(k+1)^{-1}\left[z - \hat{z}(x+1|k)\right] \le \gamma \tag{6.9}$$

with the gate threshold γ set to 11.4, which corresponds to a probability of 0.99 with three degrees of freedom, and the gate size is increased by a factor of 1.5 while maintaining preliminary tracks.

The processing flow of the tracker is described in [2] as:

- Update all confirmed tracks with the closest plot to $\hat{z}(k+1|k)$ falling within the association gate defined in (6.8). If no plots are present, rate-aid the track.
- Using any remaining plots, update all preliminary tracks with the closest plot to $\hat{z}(k+1|k)$ falling within the association gate defined in (6.8). If no plots are present, rate-aid the track.
- Using any remaining plots, initiate new tracks.

As with any Kalman filter, the Kalman gain controls the bias between measurement and process noise and the choice depends on the accuracy of measurements and the motion behaviors of the targets to be tracked. The paper [2] described the use of this algorithm with a passive radar system based on a single FM radio transmitter and a single receiver in the Netherlands, demonstrating reliable tracking in real time of commercial aircraft targets over the North Sea out to ranges beyond 150 km.

6.4.2 Probability Hypothesis Density Tracking

Another approach is that adopted by Tobias and Lanterman [12], tackling the two problems of ghost excision and target state estimation using the Probability Hypothesis Density (PHD) approach, originally developed by Mahler [12]. The PHD is defined as being any function that, when integrated over any given area, specifies the expected number of targets present in the area.

They use the particle filter implementation of the update equations, in which the PHD is represented by a collection of particles and their corresponding weights. Using the same notation as [12], at time step k each particle in the filter is a vector of the form

$$\xi_i = \left[x_i \; y_i \; \dot{x}_i \; \dot{y}_i\right]^T \tag{6.10}$$

and has a weight w_i,k, where (x_i, y_i) specify the location of the particle and $(\dot{x}_i \dot{y}_i)$ specify its velocity components. As per the defining property of the PHD,

$$\tilde{N} = E\left[no.\,of\ targets\right] = \left[N_{k|k}\right]_{nearest\ integer} \tag{6.11}$$

where

$$N_{k|k} = \sum_i w_{i,k} \tag{6.12}$$

Specifically, the PHD is expected to: (1) automatically estimate the number of targets, (2) resolve ghost targets, and (3) fuse sensor (i.e., bistatic transmit-receive pair) data without the need for any explicit report-to-track association [13].

Results were presented from a simulation using three bistatic transmit-receive pairs measuring first range and then range/Doppler on two aircraft targets flying in the Washington, D.C., area. The transmitters were three local VHF FM stations and the receiver was based on that used by Lockheed Martin's Silent Sentry, located 30–50 km from the transmitters. The simulations assumed adequate target visibility, overlapping coverage, and no multipath, and calculated signal-to-noise ratios (SNRs) ranged from 12.2 dB to 32.5 dB.

In its simplest form, the simulation began by independently and randomly assigning the particles' two-dimensional position and velocity components to fall within the field of view for each transmit-receive pair. Particle weights were initially set to zero. These particles were then propagated forward in one-second steps. Birth particles with random positions and velocities were added at each time step to model new targets. One new target and hence one birth particle was assumed to appear at each step. The PHD then assigned (and updated) particle weights $w_{i,k+1}$ at each time step by incorporating range/Doppler observations, calculated probability of detection, Poisson-distributed false alarms, and a single-target likelihood function. Finally, the expected number of targets in the FOV was calculated by means of (6.11). The locations of the N expected targets were found by extracting the N highest peaks from the PHD represented by these weights.

The results of these preliminary simulations were described as encouraging. It was observed that in areas of low SNR the number of targets was overestimated. Subsequently an improved method was developed that removed the need to restrict particles to areas of high SNR, although at the expense of greater computational load [12]. This also had the effect of reducing the number of particles needed from a few thousand to a few hundred.

6.4.3 Multireceiver Passive Tracking

A third approach [14] was described by Klein and Millet of Thales Air Systems in France and has been implemented on the Thales HA-100 Homeland Alerter passive radar system. This uses both FM radio and DVB-T illuminators, exploiting the relative advantages of both (these are discussed in detail in Chapter 3).

Information from the different bistatic transmit-receive pairs may be fused in a number of ways. Each one may form a track, and the tracks may be fused, either by selecting a best track according to some criterion or by combining the tracks, weighted according to some measure of the track quality. Alternatively, a single track may be updated from the information from each transmit-receive pair. This latter approach was adopted here because it was found that in practice it was not always possible to obtain reliable tracks from each transmit-receive pair because of the variation of quality due to bistatic geometry and (in the case of FM illumination) instantaneous modulation. The tracker architecture is shown in Figure 6.9.

The equations which relate the Cartesian position (in 3D) and velocity of the target to the bistatic range R and velocity v are highly nonlinear:

$$R = \left\| \boldsymbol{x} - \boldsymbol{x}_{Tx} \right\| + \left\| \boldsymbol{x} - \boldsymbol{x}_{Rx} \right\| - \left\| \boldsymbol{x}_{Tx} - \boldsymbol{x}_{Rx} \right\| \tag{6.13}$$

$$v = \dot{R} = \frac{\boldsymbol{x} - \boldsymbol{x}_{Tx}}{\left\| \boldsymbol{x} - \boldsymbol{x}_{Tx} \right\|} + \frac{\boldsymbol{x} - \boldsymbol{x}_{Rx}}{\left\| \boldsymbol{x} - \boldsymbol{x}_{Rx} \right\|} . \boldsymbol{v} \tag{6.14}$$

where $\boldsymbol{x}$ and v are the Cartesian position and Cartesian velocity of the target, and $\boldsymbol{x}_{Tx}$ and $\boldsymbol{x}_{Rx}$ are the Cartesian positions of the transmitter and receiver. The tracker itself is a nonlinear Kalman filter.

Experimental results are presented in terms of the proportion of time for which the targets are tracked, the average track length and ground position accuracy, showing that by all three measures the fusion of data gives improved performance. This also emphasizes the complementary nature of the DVB-T and FM radio illuminators, showing that good performance can be achieved by exploiting both together.

6.5 Summary

This chapter has reviewed a number of methods of deriving target tracks from the raw detection information in a passive bistatic radar. That information for a given target may consist of TDOA, AOA, and/or echo Doppler shift from one or more than one bistatic transmit-receive pair. Two widely used ways of presenting that information are in terms of iso-range ellipses or range-Doppler plots. The intersections of iso-range ellipses from more than one transmit-receive bistatic pair result in false target ghosts, which need to be excised.

Conventional detection and target tracking algorithms are usable with passive bistatic radar, but they need to take account of the differences from conventional monostatic radar, which include the bistatic geometry, the presence of false target ghosts, and (with some kinds of illuminator) the coarse range resolution and the dependence of quality on the bistatic geometry and on the waveform modulation. As has been seen, passive radar presents some new challenges for accurate tracking due to the coarse angular resolutions (and range resolution in the case of VHF signals) but also has some advantages, particularly in terms of higher update rates.

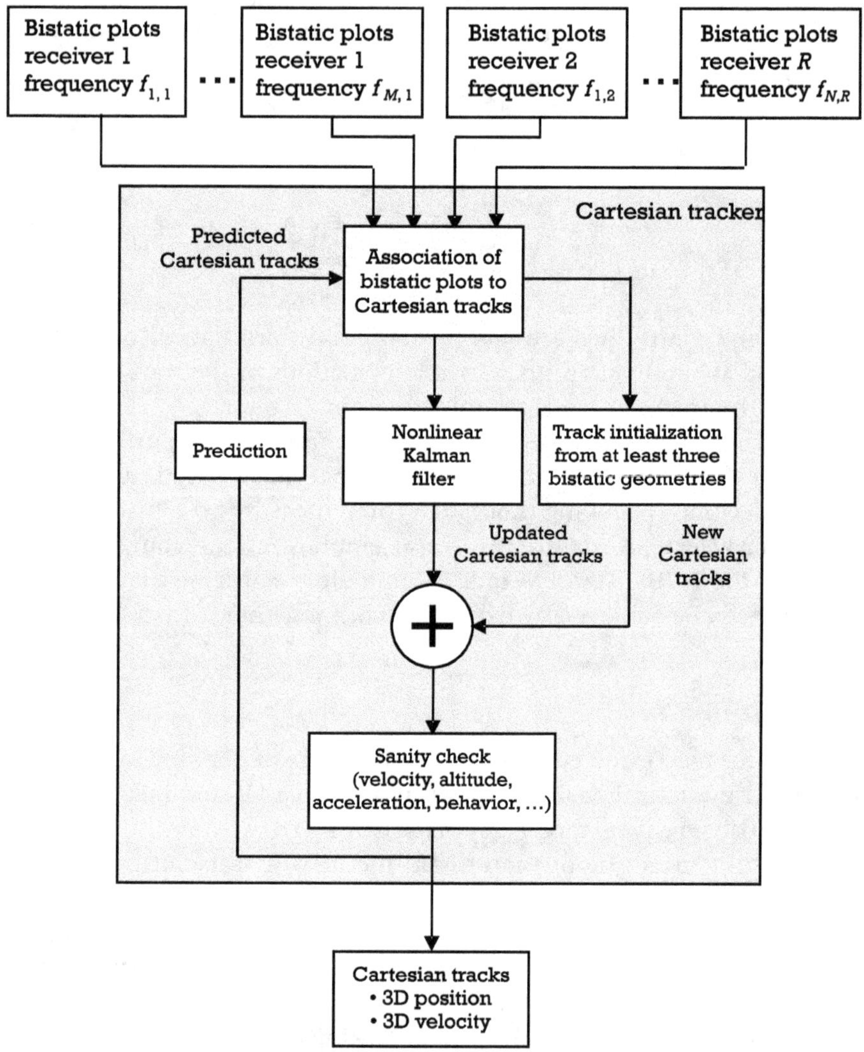

Figure 6.9 Tracking architecture used in the Thales HA-100 system. The tracker uses the plots from each bistatic transmit-receiver pair. (© 2012 IEEE. Reprinted with permission from [14].)

Three approaches to target tracking have been presented, one based on the Kalman filter, one based on the PHD algorithm, which has the additional advantage of excising false target ghosts, and one based on fusing information from several FM radio and DVB-T bistatic transmit-receive pairs.

References

[1] Richards, M. A., W. A. Holm, and J. A. Scheer, *Principles of Modern Radar: Vol. 1, Basic Principles*, Raleigh, NC: SciTech Publishing, 2010.

[2] Howland, P. E., D. Maksimiuk, and G. Reitsma, "FM Radio Based Bistatic Radar," *IEE Proc. Radar, Sonar and Navigation*, Vol. 152, No. 3, June 2005, pp. 107–115.

[3] Caspers, J. M., "Bistatic and Multistatic Radar," Ch. 36 in *Radar Handbook*, 1st ed., M. I. Skolnik, (ed.), New York: McGraw-Hill, 1970.

[4] Brown, J., et al., "Passive Bistatic Radar Location Experiments from an Airborne Platform," *IEEE AES Magazine*, Vol. 27, No. 11, November 2012, pp. 50–55.

[5] Malanowski, M., and K. Kulpa, "Two Methods for Target Localization in Multistatic Passive Radar," *IEEE Trans. on Aerospace and Electronics Systems*, Vol. 48, No. 1, January 2012, pp. 572–580.

[6] O'Hagan, D., "Passive Bistatic Radar Performance Using FM Radio Illuminators of Opportunity," Ph.D. thesis, University College London, March 2009.

[7] Brookner, E., *Tracking and Kalman Filtering Made Easy*, New York: Wiley, 1988.

[8] Blackman, S., and R. Popoli, *Design and Analysis of Modern Tracking Systems*, Norwood, MA: Artech House, 1999.

[9] Bar-Shalom, Y., X. Rong Li, and T. Kirubarajan, *Estimation with Applications to Tracking and Navigation: Theory, Algorithms and Software*, New York: Wiley, 2001.

[10] Ristic, B., S. Arulampalam, and N. Gordon, *Beyond the Kalman Filter: Particle Filters for Tracking Applications*, Norwood, MA: Artech House, 2004.

[11] Tobias, M., and A. D. Lanterman, "Probability Hypothesis Density-Based Multitarget Tracking with Bistatic Range and Doppler Observations," *IEE Proc. Radar, Sonar and Navigation*, Vol. 152, No. 3, June 2005, pp. 195–205.

[12] Mahler, R. P. S., "Multitarget Bayes Filtering Via First-Order Multitarget Moments," *IEEE Trans. on Aerospace and Electronics Systems*, Vol. 39, No. 4, October 2003, pp. 1152–1178.

[13] Tobias, M., "Probability Hypothesis Densities for Multitarget, Multisensor Tracking with Application to Passive Radar," Ph.D. thesis, Georgia Institute of Technology, 2006.

[14] Klein, M., and N. Millet, "Multireceiver Passive Radar Tracking," *IEEE AES Magazine*, Vol. 27, No. 10, October 2012, pp. 26–36.

CHAPTER

7

Contents

Examples of Systems and Results

7.1 Introduction

The purpose of this chapter is to present and discuss examples of real passive radar systems and results. These cover a wide range of illuminator sources and applications from satellite-borne transmissions, which allow synthetic aperture imaging of the Earth's surface, to indoor WiFi access points, which allow humans to be detected and tracked. The organization of the chapter essentially follows the list of emitters covered in Chapter 3. Although the number of pages available does not permit detailed descriptions, references are provided to allow the reader to follow the original publications.

7.2 Analog Television

Some of the first passive radar experiments, in the early 1980s, used analog television (TV) transmissions [1], principally because the transmissions are high-power and of reasonably broad bandwidth (~6 MHz) and the ghosting effect due to multipath that is sometimes seen with analog TV provides a simple and convincing demonstration that the technique is viable. However, it soon became apparent, as discussed in Chapter 3, that the analog TV signal is far from ideal as a radar waveform, both because of the ambiguities associated with the 64-μs line repetition rate and strong line sync pulse and because of the difficulty of suppressing the direct signal (see Chapter 4).

Better results were achieved by Howland in the mid-1990s [2]. For the reasons given above, he did not attempt to measure target range information directly, but instead used just the vision carrier part of the TV signal (shown in Figure 3.2) and extracted echo Doppler shift, and measured angle of arrival (AOA) with an interferometer formed from two Yagi antennas mounted side-by-side. This approach has been termed narrowband PBR, and the tracking of targets in this way has some similarities to passive underwater sonar, on which there is an extensive literature. Despite the lower effective transmit power associated with using just a part of the signal spectrum, the results were very positive. The processing used an extended Kalman filter, and demonstrated tracking of civil aircraft over a large part of the southeast of the United Kingdom (Figure 7.1).

7.3 FM Radio

VHF frequency modulation (FM) radio emissions have formed the basis of many experiments in passive radar, since the sources are high-power and readily available in virtually all countries worldwide. The waveform properties have been discussed in Chapter 3. Here we describe some results from typical systems.

7.3.1 Silent Sentry

In the 1990s the Lockheed Martin company in the United States developed an experimental passive radar system making use of VHF FM radio illuminators [3]. The processing exploited differential range, AOA, and Doppler information and was able (in its third version) to demonstrate reliable real-time detection and tracking of multiple air targets in the Washington, D.C., area. Claimed accuracy in tracking air targets was 100–200m (horizontal position), 1,000m (vertical position), and <2 m/s (horizontal velocity). The system also demonstrated real-time detection and tracking of rocket launches from Cape Canaveral in Florida.

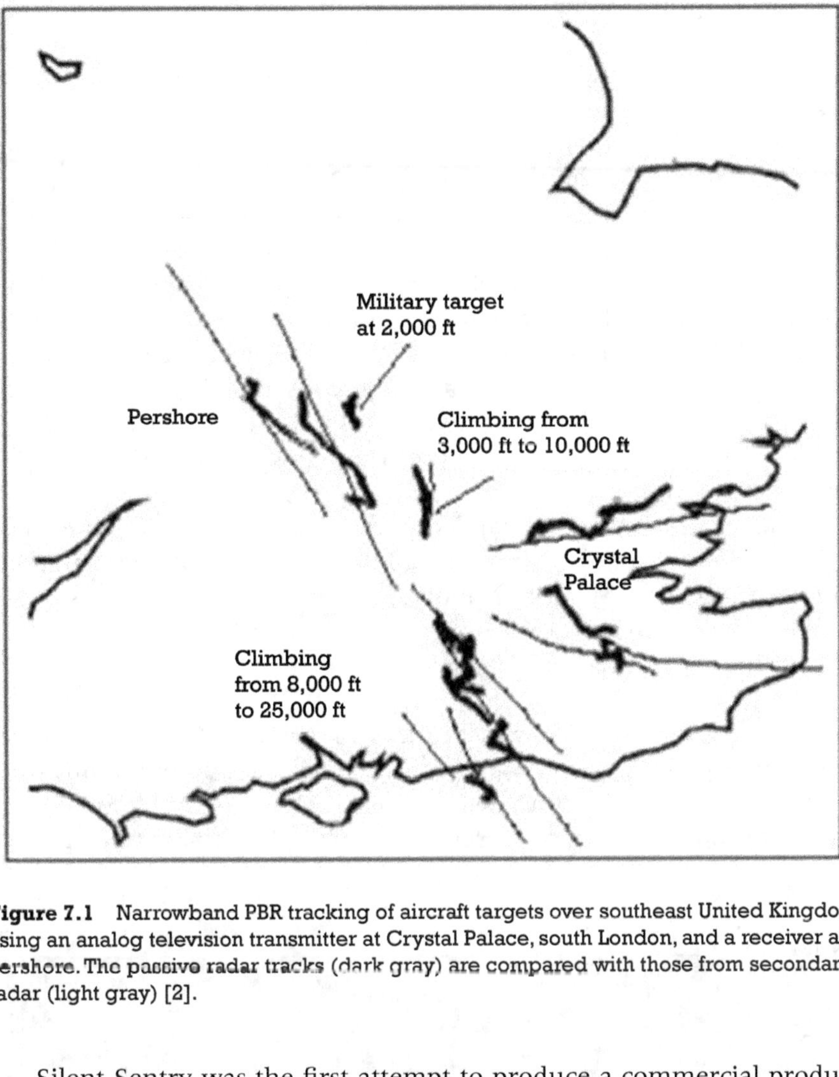

Figure 7.1 Narrowband PBR tracking of aircraft targets over southeast United Kingdom using an analog television transmitter at Crystal Palace, south London, and a receiver at Pershore. The passive radar tracks (dark gray) are compared with those from secondary radar (light gray) [2].

Silent Sentry was the first attempt to produce a commercial product based on passive radar, but it was ahead of its time, in that the killer applications for passive radar had not yet been clearly established, and the performance, while certainly impressive, was not as good as conventional radars.

7.3.2 The Manastash Ridge Radar

The Manastash Ridge Radar (MRR) was mentioned in Chapter 1. It was conceived and built also in the late 1990s by John Sahr and Frank Lind at the University of Washington, Seattle, as a low-cost approach to the study of plasma turbulence in the E-region of the ionosphere at northern latitudes [4, 5]. It used the signal from an FM radio transmitter at 96.5 MHz located in Seattle, Washington, and it solved the problem of direct signal suppression (Chapter 4) by siting the receiver remotely, the other side of the Cascade Mountains at a range of 150 km from the transmitter, so the direct signal was negligible. Synchronization and data transfer were achieved via the Internet.

Figure 7.2 shows an example of a range-Doppler plot formed from a 10-second average of received echoes out to a range of 1,200 km, with Doppler-shifted regions of E-region turbulence at ranges of about 900 and 1,050 km. The scatter at ranges near 70 km is associated with ground clutter; the largest ground clutter signal corresponds to scatter from Mount Rainier, a prominent volcano whose peak is nearly 3,000m above the surrounding terrain.

Figure 7.3 shows an expanded portion of the range-Doppler plot. Eight separate aircraft can be observed in the region ±303 m/s and 50–150-km range. The Manastash Ridge Radar was not in any sense optimized to detect aircraft targets, but they appeared quite regularly. They lie near the ground clutter detection, and no effort was made to remove them.

The MRR used simple direct-digitization receivers. Data in the format shown in Figure 7.2 was made available online on a continuous 24/7 basis. Overall, the system provided a remarkably simple and low-cost approach to ionospheric remote sensing, and of the versatility of passive

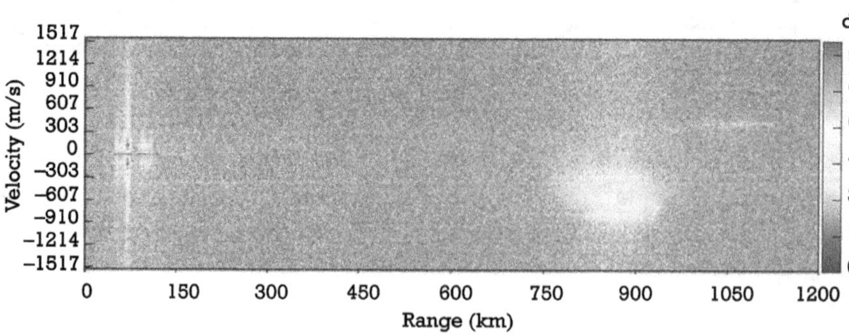

Figure 7.2 Example of a range–Doppler plot showing detection of E-region turbulence. The data were taken from 10 seconds of scatter on October 31, 2003, at 081100 UT. (Courtesy of John Sahr [5].)

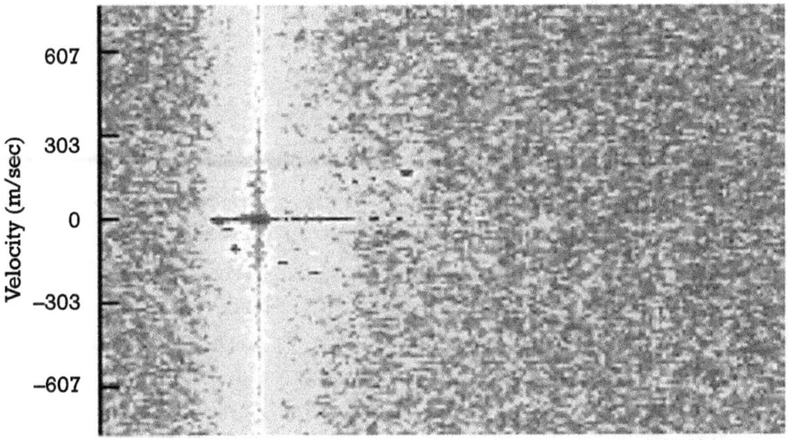

Figure 7.3 The detection signatures of eight separate aircraft can be observed in the region ±303 m/s and 50–150-km range–Doppler plot. (Courtesy of John Sahr [5].)

radar. Subsequent work has investigated the use of higher-performance receivers based on software-defined radio modules, and further comments on this subject appear in Chapter 8.

7.3.3 More Recent Experiments Using FM Radio Illuminators

Numerous VHF FM radio-based passive radar systems have been built and reported in the past two decades. Many of the publications concentrate on the difficult aspects of the technique, such as direct signal suppression or tracking algorithms, and these have been described in Chapters 4 and 6, respectively. Howland et al. [6] built and demonstrated a system using a single FM radio transmitter and a single receiver, detecting and tracking commercial airliners over the North Sea out to ranges in excess of 150 km, and the Kalman filter tracking processing used in that system has been described in Chapter 6.

The example shown in Figure 6.1 in Chapter 6 is from a system built and demonstrated by the University of Cape Town and the South African government research laboratory CSIR in the region of Pretoria, South Africa, also using a single FM radio transmitter and a single receiver, and shows broadly comparable performance.

Results of this kind have been analyzed by Malanowski and his coworkers at Warsaw University of Technology, taking into account factors such as realistic antenna radiation patterns, target RCS, and integration times (and hence processing gain) [7]. They reported experimental results with civil airliner targets using their PaRaDe equipment, validated with ADS-B tracks, at ranges in excess of 600 km, although they emphasized the importance of adequate receiver dynamic range.

7.3.4 Summary

Despite the poor range resolution and time-varying waveform properties, VHF FM signals have the significant advantages that they are plentiful and usually high-power, and have formed the basis of numerous passive radar experiments worldwide. FM radio transmissions may be expected to continue in many countries for the foreseeable future although, as noted in Chapter 1, it has been reported that Norway will start to discontinue FM radio transmissions from January 2017 and this will be computed by the end of 2017.

7.4 Cell Phone Base Stations

Cell phone base stations are ubiquitous, even in the developing world, and provide signals that can readily be exploited for passive radar. The first experiments using such signals were made in the United Kingdom in the early 2000s by the Roke Manor Laboratories, and the concept was known as Celldar [8]. However, although a number of claims were made in press releases, nothing was published in the peer-reviewed technical literature.

The radiation pattern of cell phone base station transmitters are typically arranged in 120° sectors, with the vertical-plane radiation pattern tailored in the same manner as Figure 3.16 to avoid wasting power above the horizontal. The trend, particularly in cities, is towards smaller cells with lower transmit powers. A generous upper value for the effective isotropic radiated power (EIRP) towards an air target might therefore be +20 dBW. Radar equation calculations of the kind presented in Chapter 4, making appropriate assumptions about target radar cross-section, integration gain, and receive antenna gain, show that detection ranges for air targets of a few kilometers at most would be achievable, which limits the usefulness to specific short-range applications.

Experimental work in Singapore using GSM transmissions confirmed these predictions [9]. The authors reported detection and tracking of large vehicle targets at ranges up to 1 km, and of human targets up to about 100m. The range resolution is such that targets at such ranges are not resolvable in range, but the Doppler information allows them to be isolated and identified.

Better results may be obtained with higher-power transmissions, greater receive antenna gain and longer integration times. Work at the Fraunhofer FKIE Institute in Germany has used a multiple-beam, multiple-channel array antenna with sufficient gain that the system can exploit the transmissions from several base stations, and hence track targets over a substantial area at ranges of up to 40 km from the receiver [10]. The main system parameters are given in Table 7.1. The base station transmissions in each case have a power of 10W radiated over 120° azimuth sectors, giving 100W (+20-dBW) EIRP. Figure 7.4 shows the cumulative probability of detection P_D overlaid on a map of the area for this system, as well as the locations and beam directions of the four base stations used in this calculation.

Although this performance is certainly more impressive than earlier experiments with GSM illuminators, the additional cost and complexity associated with the multi-element array and multiple receiver channels mean that this is no longer a simple system. Nevertheless, it does illustrate the potential of combining information from several illuminators over an extended area.

7.5 DVB-T and DAB

More recently, there has been a substantial shift to exploiting digital transmissions. This is driven partly by availability, especially going into the future, and partly by aspects of the waveforms that exhibit more favorable ambiguity function diagram properties (as seen in Chapter 3). In addition, there has been a much more significant involvement of industry as well as research labs. This is a sign of a maturing technology and one that has genuine commercial potential. Pressure on the areas of spectrum

Table 7.1
Experimental Parameters

Transmit power	10W
Transmit antenna gain (120° sector)	10 dB
Signal bandwidth	81.3 kHz
Receive antenna gain	25 dB
Coherent integration time	0.34 second
Processing gain	41.2 dB

Source: [10].

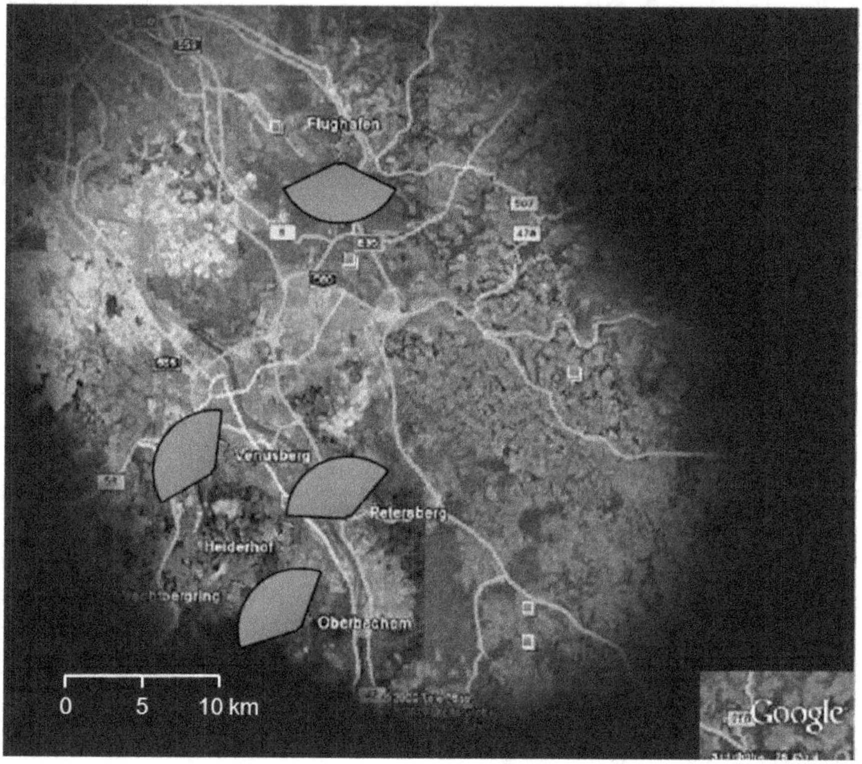

Figure 7.4 Coverage of the FKIE passive radar system using GSM illuminators, with cumulative probability of detection P_D overlaid. Four of the seven base stations are shown, each with their 120° illumination sector [10].

currently allocated to radar is likely to lead to a persistent development of passive radar that exploits digital waveforms. However, it should be borne in mind that there is no need to choose between the two. Both analog and digital signals can be readily exploited and to a degree have complementary properties.

The first publication on the use of digital radio and TV transmissions for passive radar purposes was by Poullin [11]. The properties of these waveforms and the OFDM modulation techniques on which they are based have been described in Section 3.3 of Chapter 3.

Notable work has also been undertaken in Australia with pioneering investigations of the use of digital transmissions as part of a larger Illuminator of Opportunity (IOO) project. While there have been numerous research publications that show a deep understanding of the topic, the work has been largely confined to that of research and development with no moves towards production systems.

An example of the new generation of passive radars is the system developed by engineers at Airbus Defence and Space (formerly CASSIDIAN) and is depicted in Figure 7.5 [12, 13]. This uses a combination of VHF FM radio, DAB and DVB-T illuminators, and is described by the authors as a "near-production stage multi-band mobile passive radar system."

The system architecture, showing the receivers covering the three bands and the fusion and tracking processing, is depicted in Figure 7.6. They report results from early 2013 using eight FM, approximately five DAB and three DVB-T illuminators and a cooperative target flying a loop trajectory, and changing height at each circuit. The target position was tracked to an accuracy of 30m.

Passive radar technology is actively being developed in other countries, including China, Russia, Iran, and Israel. Academic publications on passive radar from workers in these countries appear regularly in journals and conference proceedings (for example, [14–19]). Information on in-service systems can be found on various defense analysis Web sites, although these should not be regarded as completely reliable.

China is reported to have developed a new passive radar system, the DWL002 that appears to operate in the VHF band. As seen in earlier chapters, if high-power transmitters can be exploited the bistatic detection range can exceed 500 km. Use of VHF is likely to have been chosen as a counter to stealth technology, which, coupled with significant bistatic angles would potentially result in higher system sensitivity than would be possible using a conventional monostatic system operating at microwave frequencies. Figure 7.7 shows an example receiver mounted on a telescopic pole.

The Iranian ALIM system (shown in Figure 7.5) was reportedly first seen in 2011 in a parade of Iranian armed forces. It is claimed to have a maximum range of between 250 and 300 km and is able to detect slow, low-flying targets with relative ease [20]. It also operates in the VHF part of the spectrum, again suggesting an antistealth capability. It is thought to be manufactured in Iran, although it appears to have origins that come from Russia.

Indeed, the development of passive radar systems for military applications is, unsurprisingly, shrouded in secrecy. A feature of passive radar is its close relationship to more conventional emitter location techniques that are not radars at all, but only listen for transmissions. As might be

(a)

(b)

Figure 7.5 5 Examples of the new generation of passive radar systems. Clockwise from top left: GAMMA (courtesy of FKIE, Germany), HOMELAND ALERTER (courtesy of THALES, France), AULOS (courtesy of Leonardo-Finmeccanica S.p.A., Italy), ALIM (Iran), Airbus Defence and Space passive radar, SILENT GUARD (courtesy of ERA, Czech Republic).

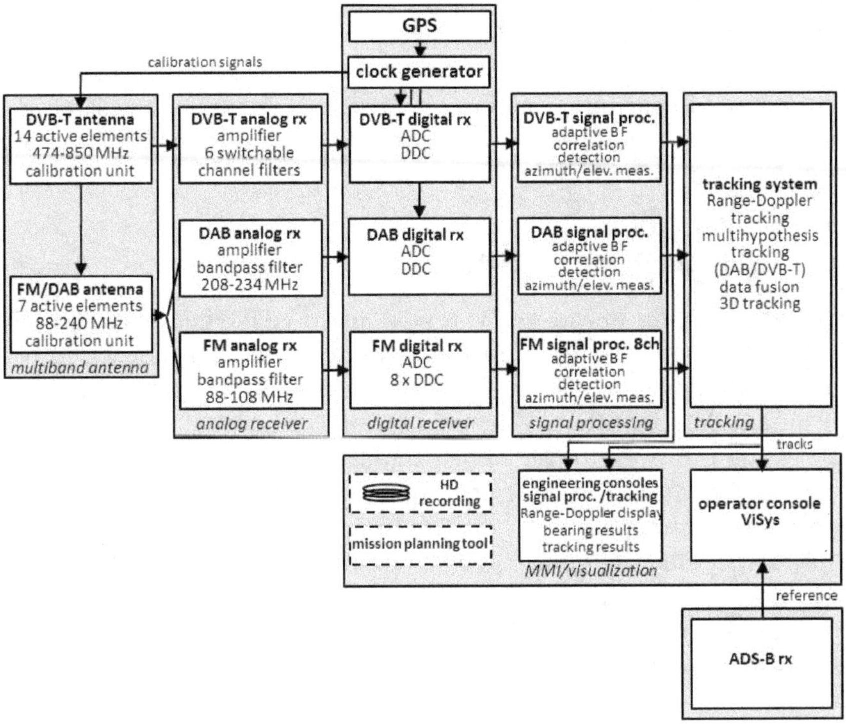

Figure 7.6 Processing architecture of the multi-band passive radar system developed by Airbus Defence and Space [13].

Figure 7.7 The Chinese DLW002 passive radar system.

imagined, the ability to exploit signals in this direct way and in the more indirect manner of passive radar suggest strongly that the two can be intertwined and used together. In the form of an extensive network, with the potential for acting as a counter to stealth technology, it is easy to understand the high levels of development that are ongoing in various countries around the world.

7.6 Airborne Passive Radar

By far, the majority of passive radar work has used fixed terrestrial receivers. However, it is also of interest to consider an aircraft-borne receiver, potentially allowing modes such as GMTI, SAR, and ISAR, or even passive bistatic Airborne Early Warning (AEW). Operationally, a stealthy aircraft would not want to use an active radar, emitting a signal that could betray its presence, so bistatic operation, including passive bistatic radar, becomes very attractive.

The first reference to airborne passive radar experiments was in 1996 [21]. More recent experiments of this kind were carried out using VHF FM illuminators and a simple multichannel receiver [22, 23]. The antenna was taped to the inside of the window of the aircraft (Figure 7.8), and the data gathered allowed iso-range ellipses to be plotted for several

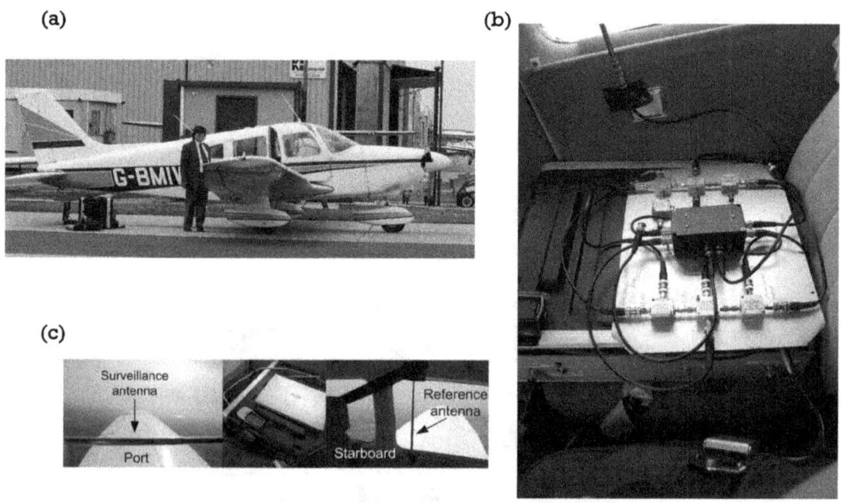

Figure 7.8 Antenna and receiver mounted inside the Piper PA 28-181 aircraft for airborne passive radar experiments (© 2010 IET. Reprinted with permission from [23].)

transmitters and targets, including target velocity vectors derived from the measured Doppler and the known velocity of the aircraft carrying the receiver. This information allowed the ambiguities associated with the different iso-range ellipses to be resolved (see Section 6.2 and Figure 6.4 in Chapter 6).

An important factor here is the vertical-plane coverage of the illuminator, especially against short-range airborne targets. This effect has been discussed in Section 3.5 of Chapter 3.

Kulpa and his coworkers at the Warsaw University of Technology in Poland carried out similar experiments, and provided a prescient assessment of the utility of airborne passive radar against ground-based stationary targets, ground-based moving targets, and airborne moving targets [24]. Their first experiments used a vehicle-borne passive receiver, which allowed specific features of the received signal, such as the Doppler spread of clutter, to be observed and quantified. The next stage in the work used a receiver mounted in a Skytruck aircraft (Figure 7.9). Their receiving system is known as PaRaDe (Passive Radar Demonstrator).

Figure 7.9 Skytruck aircraft used by Warsaw University of Technology for airborne passive radar experiments. The inset shows the antenna taped to the inside of the window [24].

A third example of airborne passive radar is tomographic imaging, using a DVB-T transmitter and an aircraft-borne receiver flying a spiral path (Figure 7.10), to give a three-dimensional (3-D) image of the target scene [25–27]. The aircraft was a venerable Cessna 170, and the receiver system used two Ettus Corp. N200 Software Defined Radio (SDR) units for the signal channel, reference (direct signal) channel and GPS (Figure 7.11). The target scene was a rural area, including some buildings and a grain silo.

The tomographic processing used to reconstruct the 3-D image is described in [25, 27] (Figure 7.12).

It is reported that the first flight of a French airborne passive radar system took place at the Salon de Provence air base in southeast France in October 2015, as part of a collaboration among the Centre de Recherche de l'Ecole de l'Air (CReA), ONERA, and SONDRA.

These experiments have only started to scratch the surface of the potential of airborne passive radar, and future work may be expected to address GMTI, STAP, and other capabilities.

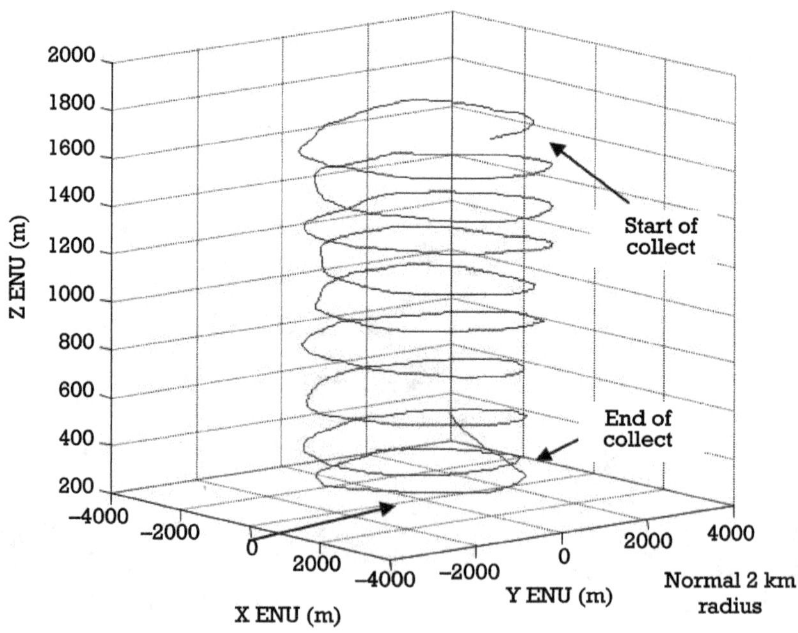

Figure 7.10 Aircraft-borne tomographic passive radar imaging. Collection aperture as-flown duration 2,550 seconds [26, 27].

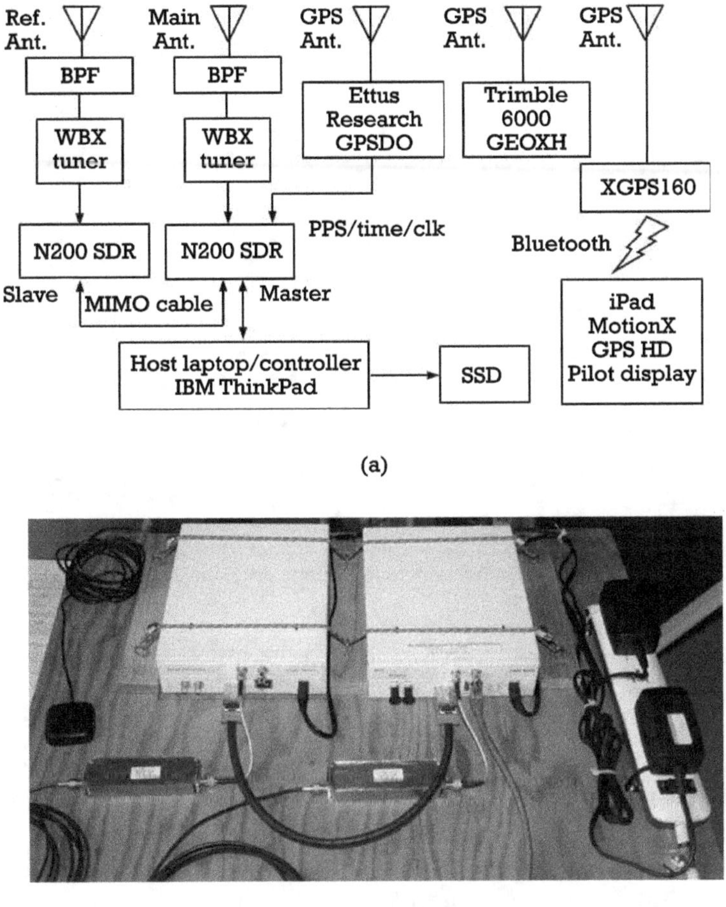

Figure 7.11 Receiver hardware [26, 27].

7.7 HF Skywave Transmissions

Yet another class of illuminator is provided by signals in the high-frequency (HF) band (2–30 MHz), propagating via reflection from the ionosphere to ranges of 1,000 km or more. These may take the form of broadcast transmissions such as the BBC World Service or Voice of America, or HF over-the-horizon radars (OTHRs) such as the Australian JORN system. The bandwidths of such signals are relatively narrow, so the radar range resolution is quite coarse, and because of the low frequency even large

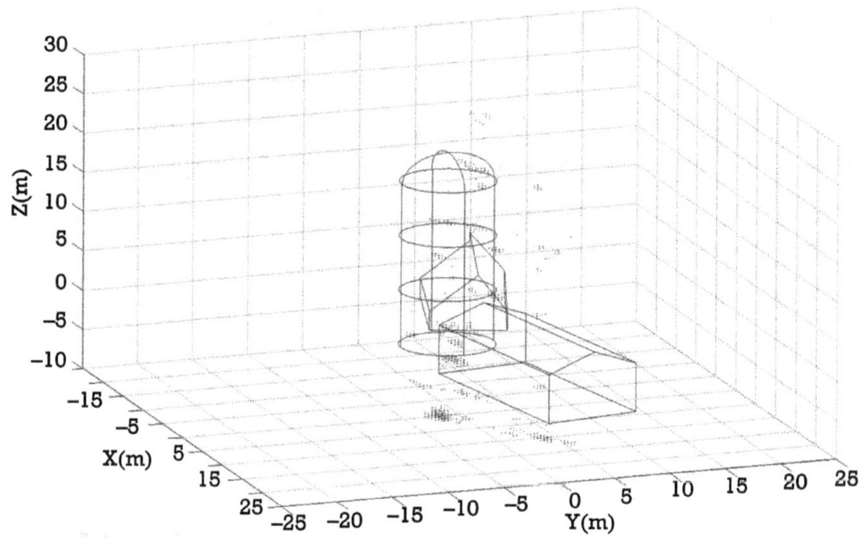

Figure 7.12 3-D tomographic image with buildings overlain [27].

antenna arrays have relatively broad beamwidth, so the azimuth resolution at such long range is also quite coarse.

Skywave propagation depends on the reflective properties of the layers in the ionosphere, which are continuously-varying and depend on the time of day, time of year, and state of the sunspot cycle, as the molecules in the atmosphere dissociate during the daytime under the influence of solar radiation and recombine at nighttime. It is the free electrons that reflect the RF signals, and the maximum frequency at which reflection occurs depends on the free electron density [28].

This kind of passive radar is suited to detection of aircraft or missile targets using a remote transmitter and receiver located closer to the target scene [29]. Lesturgie and Poullin reported the results of some experiments using a noncooperative HF broadcast transmitter in Kiev, Ukraine, and a shipborne receiver off the west coast of France some 3,000 km away, demonstrating detection of an aircraft target at a range of up to 200 km [30]. The system was called Nostramarine, and the concept is depicted in Figure 7.13.

As described in Chapter 3, digital modulation formats such as DRM (Figure 3.12) give better performance than analog, because the ambiguity function is not time-varying and does not depend on the instantaneous modulation [31–33].

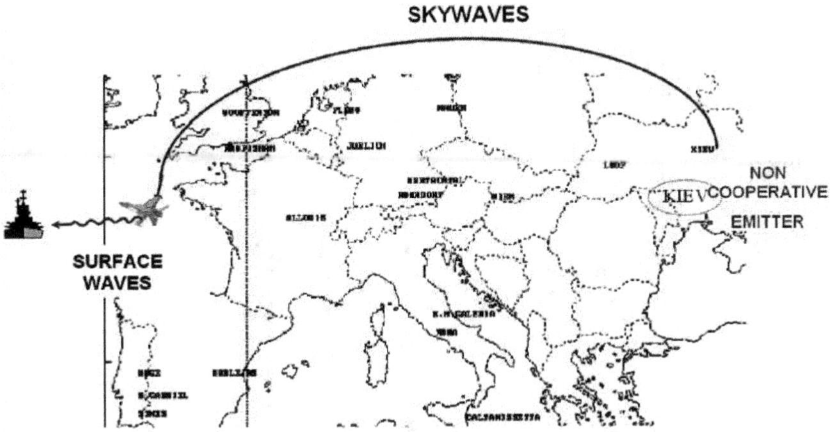

Figure 7.13 Nostramarine concept, using a noncooperative remote HF broadcast illuminator and a shipborne receiver [30].

7.8 Indoor/WiFi

The modulation formats of the IEEE Std 802.11 WiFi and IEEE Std 802.16 WiMAX signals have been described in Chapter 3. The 802.11 WiFi standard has been shown to be suitable for indoor use, in applications such as intruder detection and monitoring. One of the first demonstrations (Figure 7.14) shows a configuration with two receivers: one to receive the direct (reference) signal and one to receive target echoes. The range resolution ($c/2B \sim 25$m) is not adequate to resolve targets in range, but the Doppler-shifted echoes of human targets walking along the corridor are readily detectable [34–36].

WiFi-based passive radar has also been investigated as a sensor for airport security, as part of a European Union funded project ATOM (Airport detection and Tracking Of dangerous Materials by active and passive sensors arrays) [37]. More recently, the same kind of technology has been demonstrated as a means to provide low-cost, short-range surveillance at small private airfields [38], showing that small aircraft and paragliders, as well as human targets, can be detected and tracked at useful ranges.

Other developments of WiFi-based passive radar may allow through-wall detection of intruders or hostages [36]. Compared to a conventional through-wall radar, the use of WiFi access point within a room means that the target echoes are subject only to the one-way propagation loss through the wall. Another application, discussed in Chapter 8, is for

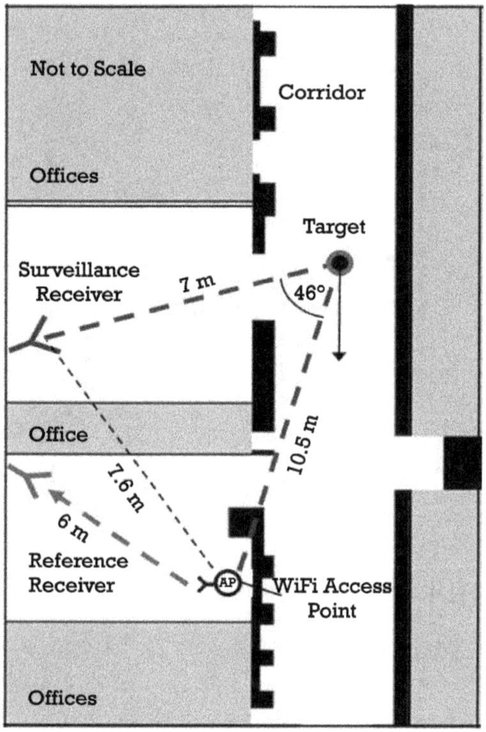

Figure 7.14 Schematic of the indoor dense clutter WiFi radar experimental setup [35].

eldercare/assisted living monitoring, so the radar echo from an individual who has fallen or is in difficulty may be distinct from that of an individual walking normally [39]. The use of radar in this application, rather than video monitoring, has advantages in respect of the privacy of the individuals concerned. Further details of this are given in Chapter 8.

As described in Chapter 3, the 802.16 WiMAX standard has higher transmit power and hence can give rather broader coverage than the 802.11 WiFi transmissions, as well as favorable ambiguity functions [40–42]. Webster and coworkers at the Naval Research Laboratory in Washington, D.C., have conducted experiments exploiting signals from two WiMAX transmitter towers, one to the northwest and one to the southeast of their receiver site, each at a distance of about 3 km from the receiver site (Figure 7.15). Table 7.2 lists the principal parameters and values [42, 43].

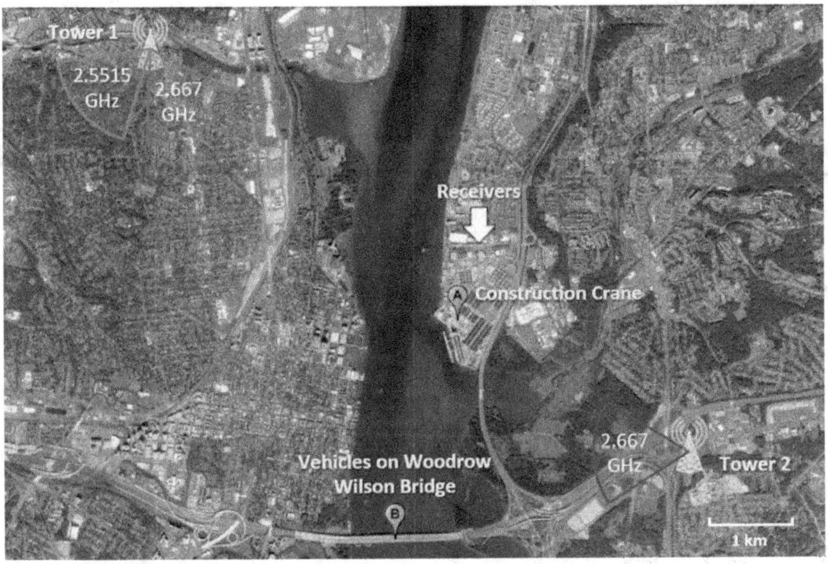

Figure 7.15 Geometry for WiMAX passive radar experiments, showing locations of transmitter towers, receiver, and targets [42].

Table 7.2
WiMAX Experiment Parameters and Values

Parameter	Symbol	Value (dB)
Transmit power	P_T	10 dBW
Transmit antenna gain	G_T	17.5 dBi
Receive antenna gain	G_R	24 dBi
Processing gain	G_P	61.76 dB
Wavelength squared	(λ^2)dB	–18.98 dBsm
Target bistatic radar cross-section	σ_B	10 dBsm
Noise power	σ_N	–127.98 dBsm

Source: [42].

Using this arrangement, they were able to demonstrate reliable detection of aircraft taking off and landing from the Washington Reagan National airport on the west side of the Potomac River, and vehicles on the Woodrow Wilson Bridge to the south.

Figure 7.16 shows detections in the form of a range-Doppler plot. Part 2 of the paper [43] showed how targets may be localized using detections from multiple bistatic transmitter-receiver pairs, by means of a technique known as multistatic velocity backprojection. This works by forming a six-dimensional data cube (position and velocity) by translating delayed and Doppler shifted data from each of the bistatic pairs within the multistatic system into a common reference frame to focus the detections.

This work demonstrates that WiMAX signals are suitable for passive radar surveillance and monitoring at ranges up to about 10 km. This scale may be appropriate for perimeter surveillance or protection of a critical asset.

7.9 Satellite-Borne Illuminators

7.9.1 Early Experiments Using GPS and Forward Scatter

As early as 1995, Koch and Westphal reported results using illumination from Global Positioning System (GPS) satellites to detect various air targets, and exploiting the forward scatter geometry (see Section 2.3) to enhance the target radar cross-section. Relatively long integration times,

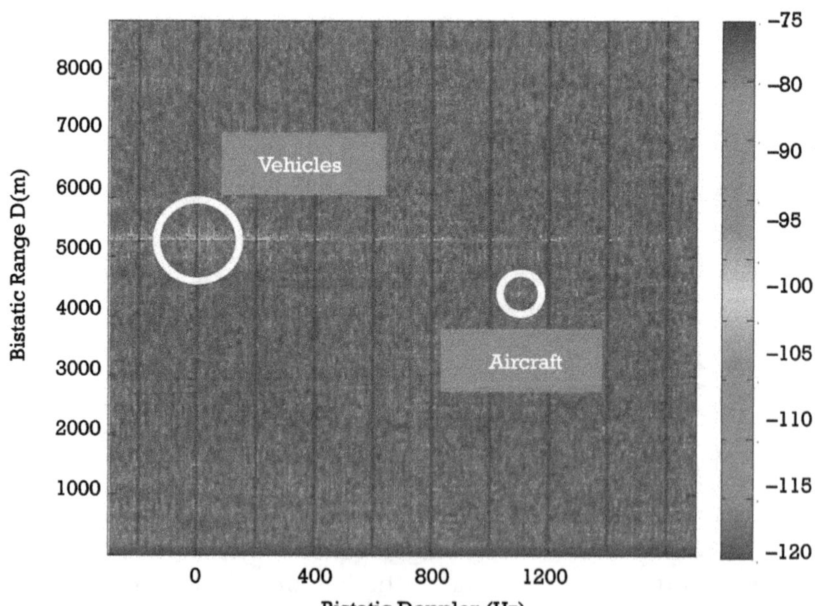

Figure 7.16 Range Doppler plot, showing detection of vehicle and aircraft targets [42].

of the order of 1 second, were used to give significant integration gain (60–70 dB). In two publications [44, 45], they reported the detection of a wide range of air targets, including civil and military aircraft, the MIR Space Station, an antitank missile, and an airship. Despite the promise shown by these results, there does not seem to have been much work done, at least in the open literature, to follow them up.

7.9.2 Geostationary Satellites

This use of the geostationary configuration in a bistatic radar with satellite TV signals was investigated and demonstrated in the early 1990s [46]. The signal power density at a target at the Earth's surface is relatively low (of the order of −107 dBW/m^2 for DBS TV; see Table 3.1), so substantial integration gain is required to give detectable target echoes at anything other than short target-to-receiver ranges. The configuration is therefore best suited to stationary target scenes, which allow long integration times.

7.9.3 Bistatic SAR

Quite soon after the first satellite-borne remote sensing synthetic aperture radars in the late 1970s, it was realized that it should be possible to use such signals as the basis of a bistatic SAR. A program COVIN REST from the mid-1980s demonstrated bistatic imaging using an aircraft-borne receiver exploiting signals from the SIR-C L-band SAR carried by the Space Shuttle, giving image resolution of the order of 20m, although the work and the results remained classified for many years [47]. Probably the first reported results in the open literature used experiments based on an aircraft-borne receiver underflying the European Space Agency's ERS-1 satellite [48], presenting an image of the Oklahoma City Airport.

Many experiments subsequently have used fixed ground-based receivers [49]. A simple consideration shows that because the echo Doppler history in such a system depends only on the variation of the one-way range (and hence phase) between the transmitter and the target (compared to the two-way range in a conventional monostatic SAR), the azimuth resolution in a bistatic SAR of this kind is equal to the length of the illuminating antenna, rather than half the length of the antenna in a monostatic stripmap-mode SAR.

Cherniakov, of the University of Birmingham in the United Kingdom, has coined the term SS-BSAR (Space-Surface Bistatic SAR) for such systems, and has demonstrated results with a number of different illuminators and configurations [50–52].

As an example, Figure 7.17 shows a bistatic SAR image of an area of Brussels, Belgium, obtained using a fixed ground-based receiver and illumination from the ASAR synthetic aperture radar carried by the European Space Agency's ENVISAT satellite, on February 15, 2016 [53, 54].

7.9.4 Bistatic ISAR

It is equally possible to use the motion of the target for synthetic aperture imaging, giving an inverse synthetic aperture radar (ISAR) [53]. Martorella, of the University of Pisa, Italy, led a set of NATO trials in 2015 to gather passive radar data from maritime targets. Experiments have used a range of different illuminators, including geostationary satellite

Figure 7.17 Bistatic SAR image (lower) and the corresponding optical image (upper) of the area to the east of the Military Hospital site, Brussels [53].

transmitters, terrestrial DVB-T transmitters and WiFi signals [56–59]. The ISAR imaging may exploit target motion through pitching or rolling.

Martorella and Giusti [60] provided a full mathematical description of the passive bistatic ISAR imaging technique, and demofnstrated it using DVB-T illumination (three adjacent channels) from a transmitter located on a hill approximately 30 km inland and a receiver located on the coast at the Naval Academy in Livorno, Italy. The processing is essentially imaging in range-Doppler space, with autofocusing to compensate for motion errors. Figure 7.18(a) shows the targets, consisting of large ships at a range of approximately 10 km from the receiver, and Figure 7.18(b) shows the focused image of the ship to the right.

7.9.5 Summary

Bistatic radar using satellite-borne illuminators has some significant attractions, including the ability to generate high-resolution SAR images with relatively simple receiver hardware. Unlike conventional space-based radar, the data is available immediately at the receiver (low latency). However, a significant disadvantage with satellites in low Earth orbit is that the illumination of the target scene is brief (only a few seconds) and repeats at the orbit repeat interval, which is typically several days.

7.10 Low-Cost Scientific Remote Sensing

Another niche application of passive radar identified in [61] is low-cost scientific remote sensing. This exploits the fact that passive radar illuminators tend to be high-power and located to give wide coverage. The relatively narrow signal bandwidth of many illuminators is not usually a problem, since many remote sensing applications do not require high spatial resolution, and hence imaging techniques. Also, suitable choice of bistatic geometry may allow an optimum regime to be found where the relationship between the radar echo and the remotely sensed quantity is monotonic and extends over a wide dynamic range. Probably the best-known example of remote sensing with passive radar is the Manastash Ridge Radar (MRR), already described in Section 7.3, but there are several other good examples.

7.10.1 Ocean Scatterometry Using GNSS Signals

A scatterometer is a radar that measures ocean surface wind speed, via the relationship between echo strength and the speed of the wind blowing

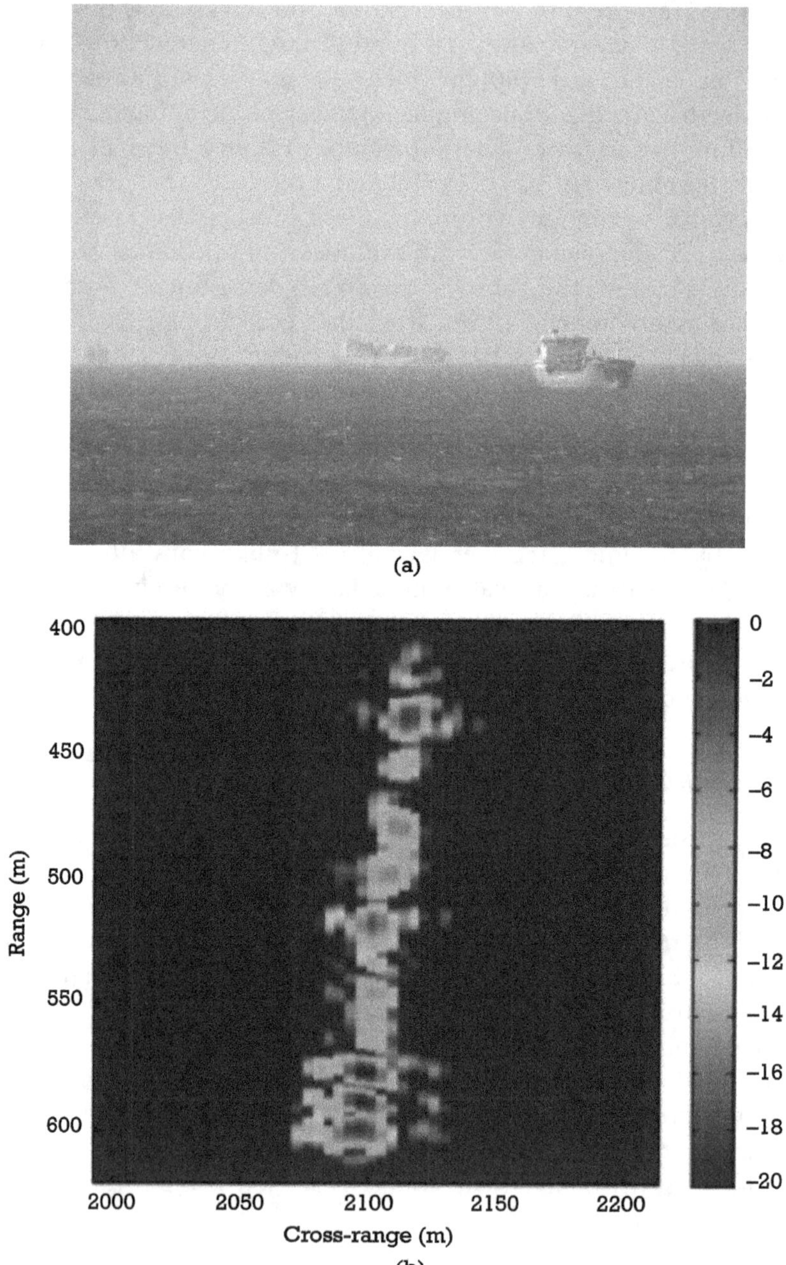

Figure 7.18 (a) Ship targets of opportunity and (b) focused ISAR image of the ship to the right. (Courtesy of Marco Martorella.)

over the ocean surface. Essentially, the higher the wind speed, the greater the wind-induced ocean surface roughness and hence the less the sea surface behaves as a specular reflector.

This idea of using GNSS signals in this way was first put forward, along with some preliminary experimental measurements with an aircraft-borne receiver, by Garrison and coworkers in 1998 [62] and came from a realization that the scattered multipath GPS signal from the sea surface that for conventional GPS purposes is a nuisance actually contains useful information about the sea surface roughness, and hence the wind speed (yet another example of the principle that "one person's interference is another person's signal"). Specifically, the width of the cross-correlation function between the scattered signal and the locally generated PRN code provides a measure of the surface wind speed. Subsequent work [63–65] carried out more detailed experiments, comparing the results with other satellite remote-sensing data and with surface truth measurements from buoys, and confirmed the viability of the technique as a simple, low-cost approach to ocean remote sensing. Signals from other GNSS systems, as listed in Chapter 3, are equally usable in this application.

7.10.2 Terrestrial Bistatic Weather Radar

The WSR-88D NEXRAD system is a network of terrestrial weather radars deployed throughout the United States to provide weather and storm information, principally as an aid to aviation. Wurman [66, 67] described an experimental bistatic receive-only adjunct to the basic network, giving improved accuracy and resolution in the recovered vector wind field. According to the definitions put forward in Chapter 1 this is a hitchhiker, because the illuminating source is an existing monostatic radar.

The receivers were known as bistatic network receivers (BNRs) and were designed to be as simple as possible, consistent with high performance. In particular, they used omnidirectional slotted-waveguide antennas, avoiding the complication and cost of a scanning antenna and pulse chasing. Particular issues that had to be taken into account included false returns via the sidelobes of the transmit antenna, and transmit-receive synchronization, which was achieved using GPS.

A total of nine BNRs were produced and deployed, in the United States, Canada, United Kingdom, Germany, and Japan, and were used in research, testing, and operations. This work demonstrates that additional passive receivers can be used to augment the performance of conventional monostatic radars in a relatively simple and cost-effective manner.

7.10.3 Planetary Radar Remote Sensing

The $1/\left(R_T^2 R_R^2\right)$ factor in the bistatic radar equation means that there are significant advantages to bistatic operation for planetary radar remote sensing if either the transmitter or the receiver can be located close to the planet being observed, and this has been recognized for many years. Simpson [68–70] identified two distinct modes of operation, shown in Figure 7.19:

- In the *uplink mode*, the illumination is provided by a high-powered transmitter on Earth with a receiver carried by a spacecraft either in orbit or flying nearby to the planet to receive and record the echo signals reflected from the planet surface. The echo information is then returned to Earth in the spacecraft telemetry stream.
- In the *downlink mode*, the illumination is provided by the spacecraft transmitter, already on board for telecommunications purposes, with Earth-based receive antennas such as those of the NASA Deep Space Network, with 70-m diameter dishes.

Table 7.3 summarizes experiments of this kind on Mars, Venus, Titan, and Pluto using Soviet, U.S., and ESA spacecraft and/or ground stations.

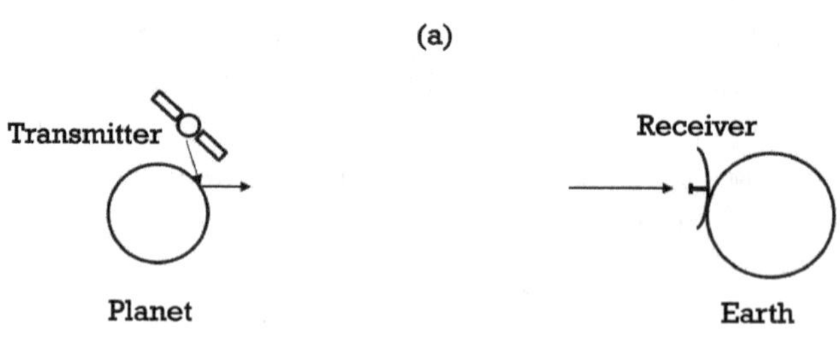

(b)

Figure 7.19 (a) Uplink mode, and (b) downlink mode.

Table 7.3
Planetary Bistatic Radar Missions and Parameters

	Spacecraft			
	Mars Express	**Venus Express**	**Cassini Orbiter**	**New Horizons**
Target	Mars	Venus	Titan	Pluto
Date	2004	2006	2006	July 2015
Mode	Downlink	Downlink	Downlink	Uplink
Bands	S, X	S, X	S, X	X
P_T **(W)**	60	5	20	10^5–10^6
G_T **(dB)**	41	26	47	73
R_T **(km)**	10,000	7,050	10,000	4.9×10^9
G_R **(dB)**	74	63	74	41
R_R (km)	1.5×10^8	1.5×10^8	1.3×10^9	60,000

Source: [70].

The first successful uplink experiment is also shown. Echoes as weak as 10^{-21} W have been detected. Additional technical and historical details, with examples of results, are reported in [68–70].

7.11 Summary

This chapter has reported a wide range of passive radar applications, systems, experiments, and results, from all over the world. The essential simplicity and low cost has meant that the subject has been very suitable for work by university groups, and at the time of this writing, radar conferences typically included several sessions devoted to passive radar. However, in more recent years, several companies have built and demonstrated systems with commercial potential. It is evident then that the subject has now come of age, so the systems and results that are reported are able to address genuine applications rather than being just academic exercises.

There is little doubt that the future for passive radar is set to be exciting. Chapter 8 considers future prospects, both in terms of new applications and new technologies.

References

[1] Griffiths, H. D., and N. R. W. Long, "Television-Based Bistatic Radar," *IEE Proc.*, Vol. 133, Pt. F, No. 7, December 1986, pp. 649–657.

[2] Howland, P. E., "Target Tracking Using Television-Based Bistatic Radar," *IEE Proc. Radar Sonar and Navigation*, Vol. 146, No. 3, June 1999, pp. 166–174.

[3] Nordwall, B. D., "Silent Sentry: A New Type of Radar," *Aviation Week and Space Technology*, Vol. 30, 1998, pp. 70–71.

[4] Sahr, J. D., and F. D. Lind, "The Manastash Ridge Radar: A Passive Bistatic Radar for Upper Atmospheric Radio Science," *Radio Science*, Vol. 32, No. 6, 1997, pp. 2345–2358.

[5] Sahr, J. D., "Passive Radar Observation of Ionospheric Turbulence," Ch. 7 in *Advances in Bistatic Radar*, N. J. Willis and H. D. Griffiths, (eds.), Raleigh, NC: SciTech Publishing, 2007.

[6] Howland, P. E., D. Maksimiuk, and G. Reitsma, "FM Radio Based Bistatic Radar," *IEE Proc. Radar, Sonar and Navigation*, Vol. 152, No. 3, June 2005, pp. 107–115.

[7] Malanowski, M., et al., "Analysis of Detection Range of FM-Based Passive Radar," *IET Radar, Sonar and Navigation*, Vol. 8, No. 2, February 2014, pp. 153–159.

[8] http://mail.blockyourid.com/~gbpprorg/mil/radar/celldar.pdf, accessed August 10, 2016.

[9] Tan, D. K. P., et al., "Passive Radar Using Global System for Mobile Communication Signal: Theory, Implementation, and Measurements," *IEE Proc. Radar, Sonar and Navigation*, Vol. 152, No. 3, June 2005, pp. 116–123.

[10] Nickel, U. R. O., "Extending Range Coverage with GSM Passive Localization by Sensor Fusion," *Proc. International Radar Symposium*, Vilnius, June 14–18, 2010.

[11] Poullin, D., "Passive Detection Using Digital Broadcasters (DAB, DVB) with COFDM Modulation," *IEE Proc. Radar, Sonar and Navigation*, Vol. 152, No. 3, June 2005, pp. 143–152.

[12] Edrich, M., and A. Schroeder, "Multiband Multistatic Passive Radar System for Airspace Surveillance: A Step Towards Mature PCL Implementations," *Proc. Int. Radar Conference RADAR 2013*, Adelaide, Australia, September 10–12, 2013, pp. 218–223.

[13] Edrich, M., A. Schroeder, and F. Meyer, "Design and Performance Evaluation of a Mature FM/DAB/DVB-T Multi-Illuminator Passive Radar System," *IET Radar, Sonar and Navigation*, Vol. 8, No. 2, February 2014, pp. 114–122.

[14] Sebt, M. A., et al., "OFDM Radar Signal Design with Optimized Ambiguity Function," *IEEE Radar Conference 2008*, Rome, Italy, May 26–29, 2008.

[15] Zaimbashi, A., M. Derakhtian, and A. Sheikhi, "GLRT-Based CFAR Detection in Passive Bistatic Radar," *IEEE Trans. on Aerospace and Electronic Systems*, Vol. 49, No. 1, January 2013, pp. 134–159.

[16] Zaimbashi, A., M. Derakhtian, and A. Sheikhi, "Invariant Target Detection in Multiband FM-Based Passive Bistatic Radar," *IEEE Trans. on Aerospace and Electronic Systems*, Vol. 50, No. 1, January 2014, pp. 720–736.

[17] You, J., et al., "Experimental Study of Polarisation Technique on Multi-FM-Based Passive Radar," *IET Radar, Sonar and Navigation*, Vol. 9, No. 7, July 2015, pp. 763–771.

[18] Yi, J., et al., "Deghosting for Target Tracking in Single Frequency Network Based Passive Radar," *IEEE Trans. on Aerospace and Electronic Systems*, Vol. 51, No. 4, October 2015, pp. 2655–2668.

[19] Yi, J., et al., "Noncooperative Registration for Multistatic Passive Radars," *IEEE Trans. on Aerospace and Electronic Systems*, Vol. 52, No. 2, April 2016, pp. 563–575.

[20] https://en.wikipedia.org/wiki/Alim_radar_system, accessed April 20, 2016.

[21] Ogrodnik, R. F., "Bistatic Laptop Radar: An Affordable, Silent Radar Alternative," *IEEE Radar Conference*, Ann Arbor, MI, May 13–16, 1996, pp. 369–373.

[22] Brown, J., et al., "Air Target Detection Using Airborne Passive Bistatic Radar," *Electronics Letters*, Vol. 46, No. 20, September 30, 2010, pp. 1396–1397.

[23] Brown, J., et al., "Passive Bistatic Radar Location Experiments from an Airborne Platform," *IEEE AES Magazine*, Vol. 27, No. 11, November 2012, pp. 50–55.

[24] Kulpa, K., et al., "The Concept of Airborne Passive Radar," *Microwaves, Radar and Remote Sensing Symposium*, Kiev, Ukraine, August 25–27, 2011, pp. 267–270.

[25] Sego, D., H. D. Griffiths, and M. C. Wicks, "Waveform and Aperture Design for Low Frequency RF Tomography," *IET Radar, Sonar and Navigation*, Vol. 5, No. 6, July 2011, pp. 686–696.

[26] Sego, D., and H. D. Griffiths, "Tomography Using Digital Broadcast TV: Flight Test and Interim Results," *IEEE Radar Conference 2016*, Philadelphia, PA, May 2–6, 2016, pp. 557–562.

[27] Sego, D., "Three-Dimensional Bistatic Tomography Using HDTV," Ph.D. thesis, University College London, September 2016.

[28] Headrick, J. M., and J. F. Thomason, "Applications of High-Frequency Radar," *Radio Science*, Vol. 33, No. 4, July–August 1998, pp. 1045–1054.

[29] Lyon, E., "Missile Attack Warning," Ch. 4 in *Advances in Bistatic Radar*, N. J. Willis and H. D. Griffiths, (eds.), Raleigh, NC: SciTech Publishing, 2007.

[30] Lesturgie, M., and D. Poullin, "Frequency Allocation in Radar: Solutions and Compromise for Low Frequency Band," *SEE Int. Radar Conference RADAR 99*, Paris, France, May 18–20, 1999.

[31] Thomas, J. M., H. D. Griffiths, and C. J. Baker, "Ambiguity Function Analysis of Digital Radio Mondiale Signals for HF Passive Bistatic Radar," *Electronics Letters*, Vol. 42, No. 25, December 7, 2006, pp. 1482–1483.

[32] Thomas, J. M., C. J. Baker, and H. D. Griffiths, "DRM Signals for HF Passive Bistatic Radar," *IET Int. Radar Conference RADAR 2007*, Edinburgh, October 15–18, 2007.

[33] Thomas, J. M., C. J. Baker, and H. D. Griffiths, "HF Passive Bistatic Radar Potential and Applications for Remote Sensing," *New Trends for Environmental Monitoring Using Passive Systems*, Hyères, France, October 14–17, 2008.

[34] Guo, H., et al., "Passive Radar Detection Using Wireless Networks," *IET Int. Radar Conference RADAR 2007*, Edinburgh, September 15–18, 2007.

[35] Chetty, K., et al., "Target Detection in High Clutter Using Passive Bistatic WiFi Radar," *IEEE Radar Conference*, Pasadena, CA, May 4–8, 2009.

[36] Colone, F., et al., "Ambiguity Function Analysis of Wireless LAN Transmissions for Passive Radar," *IEEE Trans. on Aerospace and Electronic Systems*, Vol. 47, No. 1, January 2011, pp. 240–264.

[37] Falcone, P., et al., "Active and Passive Radar Sensors for Airport Security," *2012 Tyrrhenian Workshop on Advances in Radar and Remote Sensing (TyWRRS)*, September 12–14, 2012.

[38] Martelli, T., et al., "Short-Range Passive Radar for Small Private Airports Surveillance," *EuRAD Conference 2016*, London, October 6–7, 2016.

[39] Ahmed, F., R. Narayanan, and D. Schreurs, "Application of Radar to Remote Patient Monitoring and Eldercare," *IET Radar, Sonar and Navigation*, Vol. 9, No. 2, February 2015, p. 115.

[40] Wang, Q., Y. Lu, and C. Hou, "Evaluation of WiMAX Transmission for Passive Radar Applications," *Microwave and Optical Technology Letters*, Vol. 52, No. 7, 2010, pp. 1507–1509.

[41] Chetty, K., et al., "Passive Bistatic WiMAX Radar for Marine Surveillance," *IEEE Int. Radar Conference RADAR 2010*, Arlington, VA, May 10–14, 2010.

[42] Higgins, T., T. Webster, and E. L. Mokole, "Passive Multistatic Radar Experiment Using WiMAX Signals of Opportunity. Part 1: Signal Processing," *IET Radar, Sonar and Navigation*, Vol. 10, No. 2, February 2016, pp. 238–247.

[43] Webster, T., T. Higgins, and E. L. Mokole, "Passive Multistatic Radar Experiment Using WiMAX Signals of Opportunity. Part 2: Multistatic Velocity Backprojection," *IET Radar, Sonar and Navigation*, Vol. 10, No. 2, February 2016, pp. 238–255.

[44] Koch, V., and R. Westphal, "A New Approach to a Multistatic Passive Radar Sensor for Air Defense," *IEEE Int. Radar Conference RADAR 95*, Arlington, VA, May 8–11, 1995, pp. 22–28.

[45] Koch, V., and R. Westphal, "New Approach to a Multistatic Passive Radar Sensor for Air/Space Defense," *IEEE AES Magazine*, Vol. 10, No. 11, November 1995, pp. 24–32.

[46] Griffiths, H. D., et al., "Bistatic Radar Using Satellite-Borne Illuminators of Opportunity," *Proc. RADAR-92 Conference*, Brighton, IEE Conf. Publ. No. 365, October 12–13, 1992, pp. 276–279.

[47] Rigling, B. D., "Spotlight Synthetic Aperture Radar," Ch. 10 in *Advances in Bistatic Radar*, N. J. Willis and H. D. Griffiths, (eds.), Raleigh, NC: SciTech Publishing, 2007.

[48] Martinsek, D., and R. Goldstein, "Bistatic Radar Experiment," *Proc. EUSAR '98, European Conference on Synthetic Aperture Radar*, Berlin, Germany, 1998, pp. 31–34.

[49] Whitewood, A., C. J. Baker, and H. D. Griffiths, "Bistatic Radar Using a Spaceborne Illuminator," *IET Int. Radar Conference RADAR 2007*, Edinburgh, October 15–18, 2007.

[50] He, X., M. Cherniakov, and T. Zeng, "Signal Detectability in SS-BSAR with GNSS Non-Cooperative Transmitters," *IEE Proc. Radar, Sonar and Navigation*, Vol. 152, No. 3, June 2005, pp. 124–132.

[51] Cherniakov, M., et al., "Space-Surface Bistatic Synthetic Aperture Radar with Global Navigation Satellite System Transmitter of Opportunity: Experimental Results," *IET Radar, Sonar and Navigation*, Vol. 1, No. 6, December 2007, pp. 447–458.

[52] Antoniou, M., R. Zuo, and M. Cherniakov, "Passive Space-Surface Bistatic SAR Imaging," *7th EMRS DTC Technical Conference*, Edinburgh, July 13–14, 2010.

[53] Kubica, V., "Opportunistic Radar Imaging Using a Multichannel Receiver," Ph.D. thesis, University College London, March 2016.

[54] Kubica, V., X. Neyt, and H. D. Griffiths, "Along-Track Resolution Enhancement and Sidelobe Reduction for Bistatic SAR Imaging in Burst-Mode Operation," *IEEE Trans. on Aerospace and Electronic Systems*, Vol. 52, No. 4, August 2016, pp. 1568–1575.

[55] Chen, V. C., and M. Martorella, *Inverse Synthetic Aperture Radar Imaging: Principles, Algorithms and Applications*, Edison, NJ: IET, 2014.

[56] Colone, F., et al., "WiFi-Based Passive ISAR for High-Resolution Cross-Range Profiling of Moving Targets," *IEEE Trans. on Geoscience and Remote Sensing*, Vol. 52, No. 6, June 2014, pp. 3486–3501.

[57] Olivadese, D., et al., "Passive ISAR Imaging of Ships Using DBV-T Signals," *IET Int. Radar Conference 2012*, Glasgow, U.K., October 2012, pp. 64–68.

[58] Turin, F., and D. Pastina, "Multistatic Passive ISAR Based on Geostationary Satellites for Coastal Surveillance," *2013 IEEE Radar Conference*, Ottawa, Canada, May 2013.

[59] Pastina, D., M. Sedehi, and D. Cristallini, "Passive Bistatic ISAR Based on Geostationary Satellites for Coastal Surveillance," *IEEE Int. Radar Conference 2010*, Arlington, VA, May 2010, pp. 865–870.

[60] Martorella, M., and E. Giusti, "Theoretical Foundation of Passive ISAR Imaging," *IEEE Trans. on Aerospace and Electronic Systems*, Vol. 50, No. 3, July 2014, pp. 1701–1714.

[61] Willis, N. J., and H. D. Griffiths, (eds.), *Advances in Bistatic Radar*, Raleigh, NC: SciTech Publishing, 2007.

[62] Garrison, J. L., S. J. Katzberg, and M. I. Hill, "Effect of Sea Roughness on Bistatically Scattered Range Coded Signals from the Global Positioning System," *Geophys. Res. Lett.*, Vol. 25, No. 13, July 1, 1998, pp. 2257–2260.

[63] Garrison, J. L., et al., "Wind Speed Measurement Using Forward Scattered GPS Signals," *IEEE Trans. on Geoscience and Remote Sensing*, Vol. 40, No. 1, January 2002, pp. 50–65.

[64] You, H., et al., "Stochastic Voltage Model and Experimental Measurement of Ocean-Scattered GPS Signal Statistics," *IEEE Trans. on Geoscience and Remote Sensing*, Vol. 42, No. 10, October 2004, pp. 2160–2169.

[65] Garrison, J. L., et al., "Estimation of Sea Surface Roughness Effects in Microwave Radiometric Measurements of Salinity Using Reflected Global Navigation Satellite System Signals," *IEEE Geoscience and Remote Sensing Letters*, Vol. 8, No. 6, November 2011, pp. 1170–1174.

[66] Wurman, J., "Vector Winds from a Single-Transmitter Bistatic Dual-Doppler Radar Network," *Bulletin of the American Meteorological Society*, Vol. 75, No. 6, June 1994.

[67] Wurman, J., "Wind Measurements," Ch. 8 in *Advances in Bistatic Radar*, N. J. Willis and H. D. Griffiths, (eds.), Raleigh, NC: SciTech Publishing, 2007.

[68] Tyler, G. L., and R. A. Simpson, "Bistatic Radar Measurements of Topographic Variations in Lunar Surface Slopes with Explorer 35," *Radio Science*, Vol. 5, 1970, pp. 263–271.

[69] Simpson, R. A., "Spacecraft Studies of Planetary Surfaces Using Bistatic Radar," *IEEE Trans. on Geoscience and Remote Sensing*, Vol. 31, No. 2, March 1993, pp. 465–482.

[70] Simpson, R. A., "Planetary Exploration," Ch. 5 in *Advances in Bistatic Radar*, N. J. Willis and H. D. Griffiths, (eds.), Raleigh, NC: SciTech Publishing, 2007.

CHAPTER

8

Contents

Future Developments and Applications

8.1 Introduction

The previous chapter has described a wide range of applications, systems and practical results. This chapter considers some more recent and speculative applications and topics and the directions that work on this subject are likely to take over the next couple of decades.

8.2 The Spectrum Problem and Commensal Radar

8.2.1 The Spectrum Problem

A major set of issues facing all users of the electromagnetic spectrum is the ever-increasing demand for the strictly finite spectrum resource. The radio frequency (RF) spectrum is used for a wide range of purposes including communications, radio and television broadcasting, radio navigation, and sensing. In the case of communications and broadcasting, greater bandwidth is needed to satisfy the growing consumer demand for higher data rates, particularly to mobile devices (e.g., streaming video-rate data to a smartphone or tablet personal computer [1–3]).

Demands on additional radar sensing will continue to grow with new applications emerging in the civil, security, and military sectors. For example, in the civil domain, air traffic is anticipated to double by 2030. Further, with new unmanned air traffic, these predicted numbers are set to double again. This will place additional emphasis on both cooperative and noncooperative sensing in order to maintain today's current safety standards. Of course, higher bandwidth for radar translates into finer-range resolution, which directly relates to sensing capability (for example, to detect and identify an in-bound hostile target). As the demand continues to grow for more access to spectrum by all these interested parties, there will be ever-greater competition for this finite resource.

At a technology level there are several approaches. One approach is to generate spectrally cleaner waveforms (for all types of transmission), so that signals can be more closely spaced without causing interference, and digital waveform generation and adaptive compensation for the errors introduced by power amplifier stages are now becoming practicable [4]. Another approach is the use of cognitive radio techniques, which dynamically adapt the disposition of signals in frequency, direction, coding, and polarization according to the prevailing spectrum occupancy [1]. Air surveillance radar concepts that exploit narrowband waveforms have also been recently developed and successfully deployed [5]. These concepts use staring transmissions that fill an entire field of regard together with an array-based receive antenna. This allows for long integration times so that the wide bandwidths necessary for high resolution in range can be traded for narrowband waveforms and high resolution in Doppler.

8.2.2 Commensal Radar

Passive radar techniques, too, have an increasingly important role to play as the spectrum problem becomes more severe. The concept of commensal radar was mentioned in Chapter 1. Here, the modulation formats of broadcast or communications waveforms are designed so that they not only fulfill their primary purpose, but are also optimized in some sense as radar signals. There is a range of possibilities here. At one extreme, the waveform and its coverage may be completely cooperative and may even be dynamically varied to optimize its performance as a radar illuminator. At the other extreme, the waveform and its coverage may be designed to be as unfavorable as possible. It is the former case that constitutes commensal radar.

Although the idea of embedding communications or telemetry information within a radar signal has been suggested by several authors [6–8], the problem is best approached from the point of view of taking modern digital communications or broadcast signal formats of the types described in Chapter 3 and considering how they may be adapted, either in the signals themselves or in the way they are processed in a radar receiver, to give favorable ambiguity functions, with high-range and Doppler resolution and low sidelobes. Chapter 3 showed that the modulation formats [especially orthogonal frequency division multiplexing (OFDM)] are essentially noise-like, but the various pilot, preamble, and prefix signals cause unwanted sidelobe features in the ambiguity functions. Greatest effort is therefore directed at means of suppressing these.

However, as pointed out in [9], this dual-use philosophy should extend not only to the waveforms, but also to the coverage, in both azimuth and elevation. Thus, currently, the coverage of a radio or television transmitter on the coast might be optimized so as not to radiate much power out to sea or above the horizon. If it is also to be used as an illuminator for a passive radar for maritime and/or air surveillance, the radiation in these directions should deliberately be optimized. As the spectrum problem becomes more acute, it will be necessary to address these issues collaboratively.

8.3 Passive Radar in Air Traffic Management

Recently, there has been a great deal of interest from a number of the major aerospace countries in applying passive radar to the problem of remote surveillance for air traffic management applications. THALES, NATS, Roke Manor, Airbus Defence and Space (formerly Cassidian), and Leonardo (formerly SELEX-SI) all have systems in development, although none have been operationally deployed at the time of writing.

These systems can be grouped under the generic name Multi-Static, Primary Surveillance Radar (MSPSR) [10]. They use transmitters of opportunity primarily in the very high frequency (VHF) and ultrahigh frequency (UHF) bands, which thus gives potential advantages in detection range and coverage under poor weather conditions and in localization updates due to the continuous nature of the transmissions. Long-range vertical coverage remains an issue but one that can be tackled using a denser network of receiving stations. The placement of the receiving stations can be tailored to a given surveillance volume and consequently can facilitate improved coverage at lower elevations. One of the main attractions of using passive radar is the avoidance of providing a transmitter that lowers overall systems costs. Figure 8.1 shows a schematic representation of the MSPSR concept.

Figure 8.1 shows how the receivers process signals for each transmitter, extract target data, and evaluate target location through a combination of techniques that can include ellipsoid intersection points occurring from receptions across all the transmitters. The plot extracted in this way contains three-dimensional (3-D) information in both position and velocity by additionally utilizing multiperspective Doppler. The system concept can survey broader and broader volumes as it is based on the use of a number of interconnecting cells, arranged to cover the area to be controlled (e.g., Approach/TMA or En-route).

It is perfectly possible to mix passive and active radar and this is an approach that has been discussed in the research literature. However, there has been little in the way of demonstrations of specifically developed

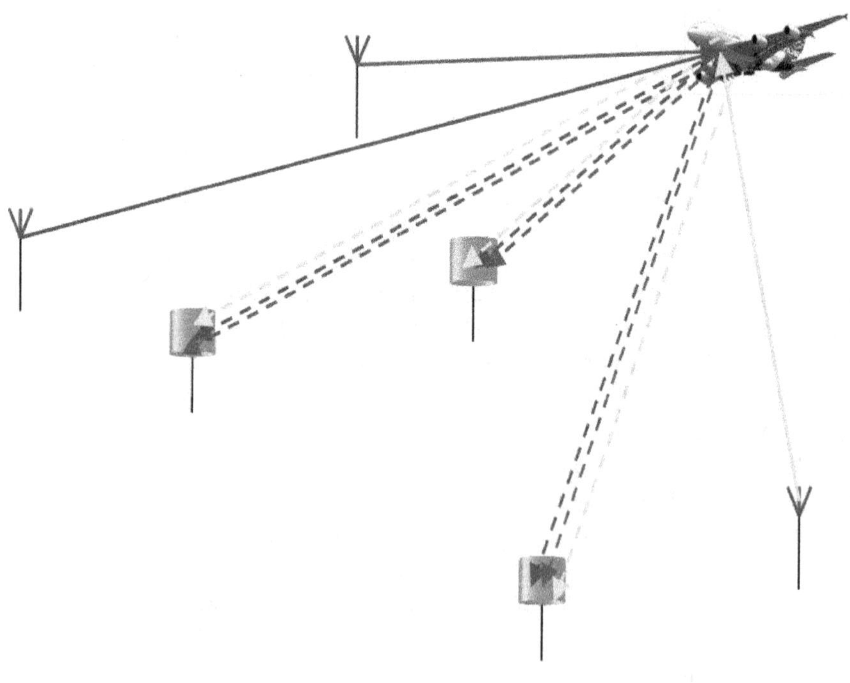

Figure 8.1 MSPSR concept.

equipment. Overall, passive radar is rapidly maturing as a viable air traffic management technology and is beginning to create considerable interest that also includes countries which currently have little in the way of air traffic management infrastructure where lower costs are especially attractive.

8.4 Countermeasures Against Passive Radar

One of the advantages of passive radar that is often quoted is that it is covert. If it is not even known whether an adversary is using passive radar techniques, clearly the deployment of countermeasures will be difficult.

8.4.1 Countermeasures

Historically, this issue was faced by the British in World War II with the German Klein Heidelberg (KH) bistatic hitchhiker system that used the British Chain Home (CH) radars as its illumination source. This was described briefly in Chapter 1. KH had been in use since mid-1943, although the British did not find out about it until October 1944. However, [11] and the minutes of three meetings held at the Air Ministry in Whitehall, London, in late 1944 and early 1945 show that no fewer than eight approaches to countermeasures against a bistatic hitchhiker of this kind were considered by British scientists. These are listed and discussed in [12], but in the context of the general problem can be summarized as:

1. Tailoring of the illuminator coverage, so as to give poor or reduced coverage of regions that would be of greatest interest from a radar point of view. The limiting case of this is to turn off the illuminator altogether for some or all of the time.

2. Modification of the illuminator waveforms, if this is feasible, to give poor ambiguity function performance, or to make it difficult to synchronise or to suppress the direct signal. Time-varying waveforms may be useful in this respect, or even switching between two or more transmitters on the same frequency.

3. Noise jamming can be used to decrease the sensitivity of receivers, and in particular to deny the reference signal to the receiver, but unless the location of the receiver is known, the jamming will

have to be spread over a wide range of angles, which will dilute its effectiveness.

4. Multiple false targets at different ranges and with different Doppler shifts can be generated, to confuse and/or overload the detection and tracking processing. In World War II the British had developed an ingenious repeater jammer called MOONSHINE, based on electro-acoustic technology [13].

More recently, Schüpbach and Böniger [14] investigated jamming techniques against DAB-based passive radar. Their strategy is to jam just the cyclic prefix part of the signal, to corrupt the reference at the beginning of each DAB frame at the receiver. This is shown to be effective and represents a very efficient use of the jamming power.

These ideas provide some indication of the types of countermeasure that may be considered against passive radar, though the details of implementation and performance are likely to be classified.

8.4.2 Bistatic Denial

Bistatic denial is a technique that has been proposed to prevent a hostile bistatic receiver from hitchhiking off a conventional radar [15]. It does this by radiating, in addition to its conventional radar signal, a masking signal to prevent the hostile bistatic receiver acquiring a reference signal from the radar [Figure 8.2(a)]. The masking signal is radiated via a radiation pattern with a null in the direction of the main beam of the radar [Figure 8.2(b)] and is coded so as to be orthogonal to the radar signal, so that the radar detector does not respond to the masking signal. The orthogonality should be maintained also as a function of Doppler.

Reference [15] considered a number of radar and masking signal coding and radiation pattern techniques and concluded that masking of the coherent reference is achievable at the same time as adequate suppression of the masking signal in the radar receiver.

8.5 Target Recognition and Passive Radar

At first sight, passive radar does not seem particularly well suited to the problem of target recognition, because the range resolution is usually too coarse. Nevertheless, passive radar echoes do contain information that can be used. Much of this comes from the high Doppler resolutions and continuously staring nature of passive radar, which results in higher

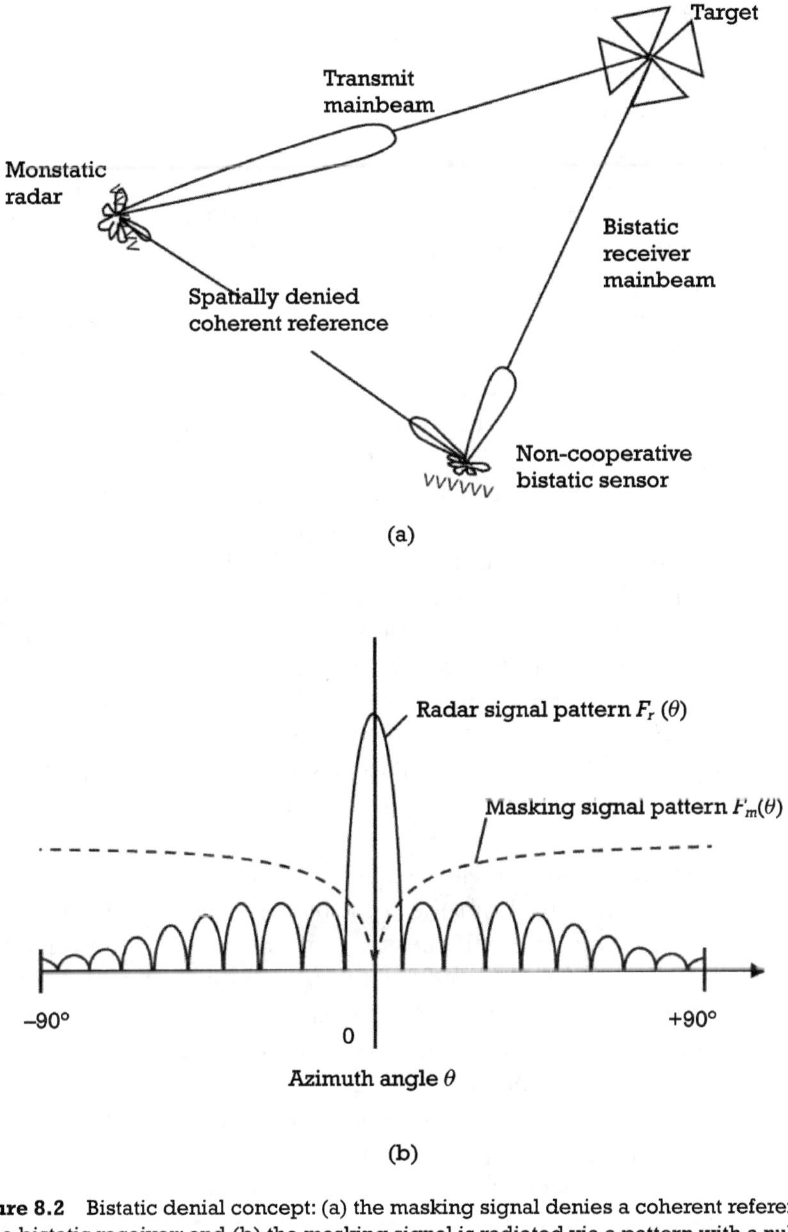

Figure 8.2 Bistatic denial concept: (a) the masking signal denies a coherent reference to the bistatic receiver; and (b) the masking signal is radiated via a pattern with a null in the direction of the radar main beam [15].

update rates and are possible using conventional scanning techniques. However, there has been reported success by Pisane et al. [16], who have used a combination of trajectory type and RCS magnitude to classify civil aircraft into broad categories.

Ehrman and Lanterman [17], Olivadese et al. [18], and Garry [19] have shown that high cross-range resolutions are possible using aperture synthesis based on the well-known ISAR technique. Here the target is allowed to traverse in a directional roughly orthogonal to the radar line of sight and successive samples are taken in time from which an aperture in the cross-range dimension can be synthesized. Figure 8.2 shows an example where a large civil airliner has been imaged using a UHF transmitter and receiver geometry such that the aircraft is imaged as it overflies the receiver to obtain the highest spatial resolution possible.

The image in Figure 8.3 represents one of the very first passive radar images, and although there is a clear correspondence with the major scatterers illustrated in the line drawing, it does not show the full details observable in many conventional high-resolution imaging radars.

A second component in the classification of air targets can be gained through the exploitation of micro-Doppler signatures. Micro-Doppler is caused by parts of a target that are in motions differently to the bulk velocity. One example of this would be the rotor components of a propeller blade as we saw in Chapter 3. Another example is the echo caused by the main and tail rotors of helicopters. Figure 8.4 shows this in the form of a range-Doppler map plus a cut along the Doppler axis at the range of the helicopter echoes.

The image in Figure 8.4 clearly shows distinct Doppler components that arise from scattering from the rotor blades. These are illustrated in the Doppler spectrum where the positions of the Doppler lines relative to one another can be extracted and used to classify the helicopter. This allows the main and tail rotor blade frequencies and number of blades (and hence rotation rate) to be computed even allowing gearbox ratios to be calculated as shown in Table 8.1.

A second way in which micro-Doppler can be exploited is through the Jet Engine Modulation (JEM) signature. The various turbine stages that comprise a typical jet engine scatter electromagnetic energy in a way that also has a very distinctive micro-Doppler signature and can provide detailed information for target classification. Figure 8.5 shows an example for a Boeing 737 civil airliner in which the main body return and the JEM lines can be easily observed through a range-Doppler map. The resulting spectrum will include components due to the differential rotations of the

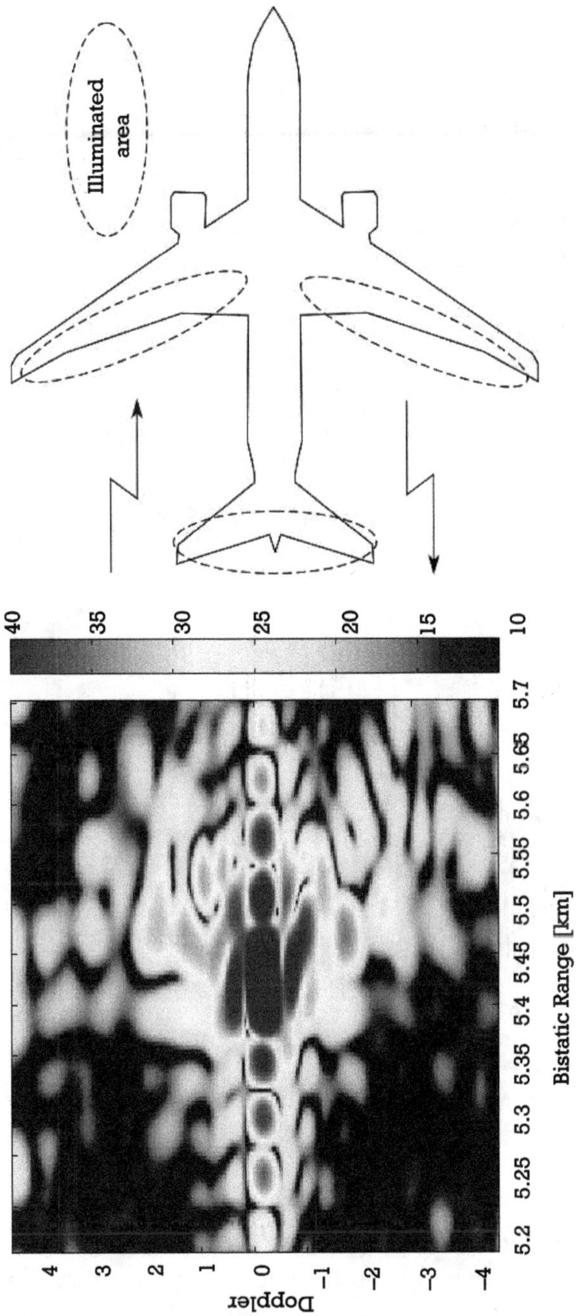

Figure 8.3 Passive radar image of civil airliner with schematic showing the relative positions of the main scatterers [19].

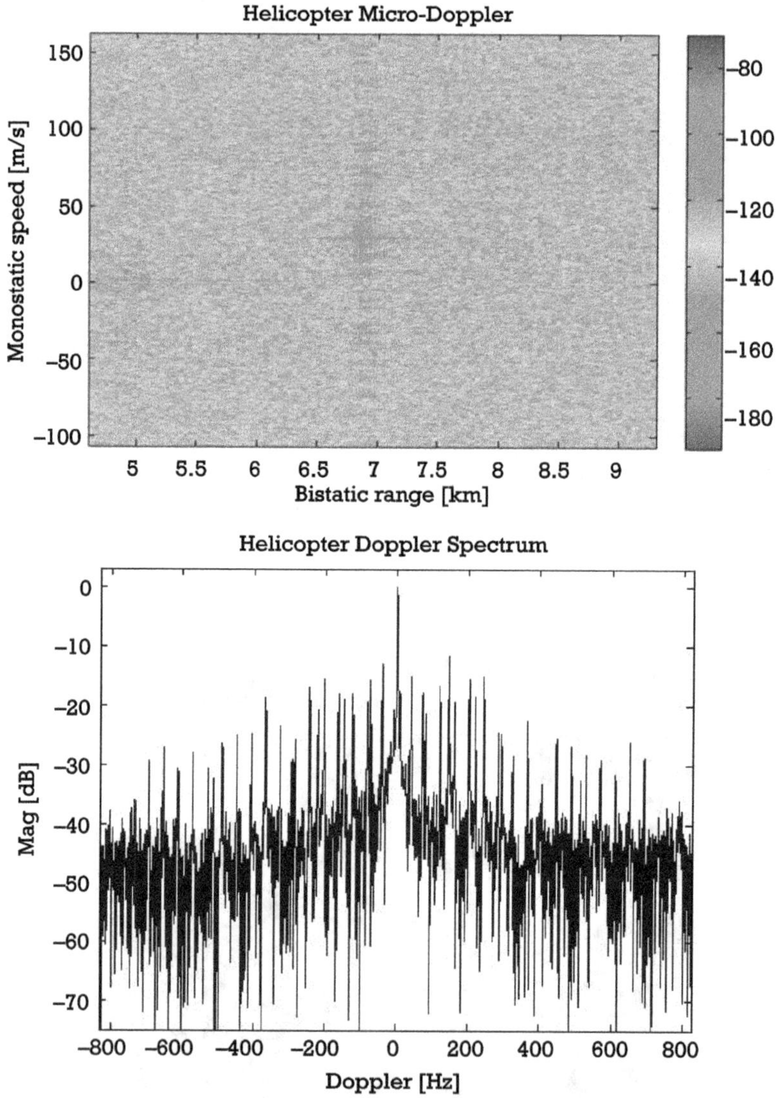

Figure 8.4 Range-Doppler map showing the micro-Doppler signature of a helicopter and corresponding Doppler lines [19].

various turbine stages together with intermodulation products that, in principle, contain detailed information for classification. By analyzing the resulting spectrum, a decision can be made as to the aircraft type [20].

These examples are merely early indicators of the potential for passive radar to be able to contribute to the complex topic of target classification

Table 8.1
Doppler, Blade Count, Blade RPM, and Gear Ratios for the Helicopter Data Shown in Figure 8.4

		Blades	RPM	Gear Ratio
Main	39.6 Hz	4	600	1:1
Tail	73.8 Hz	2	2,160	3.6:1

and there remains much further research to be completed. However, the future looks promising given that a typical passive radar system is likely to have multiple transmitters and receivers so that multiple looks and multiple frequencies can be exploited, adding further to the information that can be extracted. Whether or not this can compensate for traditional techniques that use higher frequencies and wider bandwidths remains to be seen. Alternatively, it could come about that spectrum crowding leads to wide contiguous bandwidths for passive radar, enabling both high Doppler resolution and higher-range resolution techniques to be combined.

8.6 Eldercare and Assisted Living

One of the applications for passive radar identified in Chapter 7 was the use of transmissions from WiFi access points to provide short-range indoor detection and monitoring. This idea can be further developed to help in the provision of independent living for the elderly. In this way, remote monitoring can provide detection and localization of falls, which is a significant issue for those living in homes or residences for the elderly population. Radar-based techniques have the advantages over video monitoring that they are not invasive of privacy and do not depend on particular lighting conditions. This might be particularly important in monitoring an individual in the bathroom, where the likelihood of slippage or falling might be high [21].

References [22, 23] describe early experiments of this kind, showing that the radar signature of an individual who has fallen can be distinguished from one who is moving normally. As an example, Figure 8.6 shows a measured spectrogram of a fall event obtained under laboratory conditions [23]. A sensor system of this kind could learn the pattern of behavior of an individual and could summon assistance automatically in the event of an anomaly.

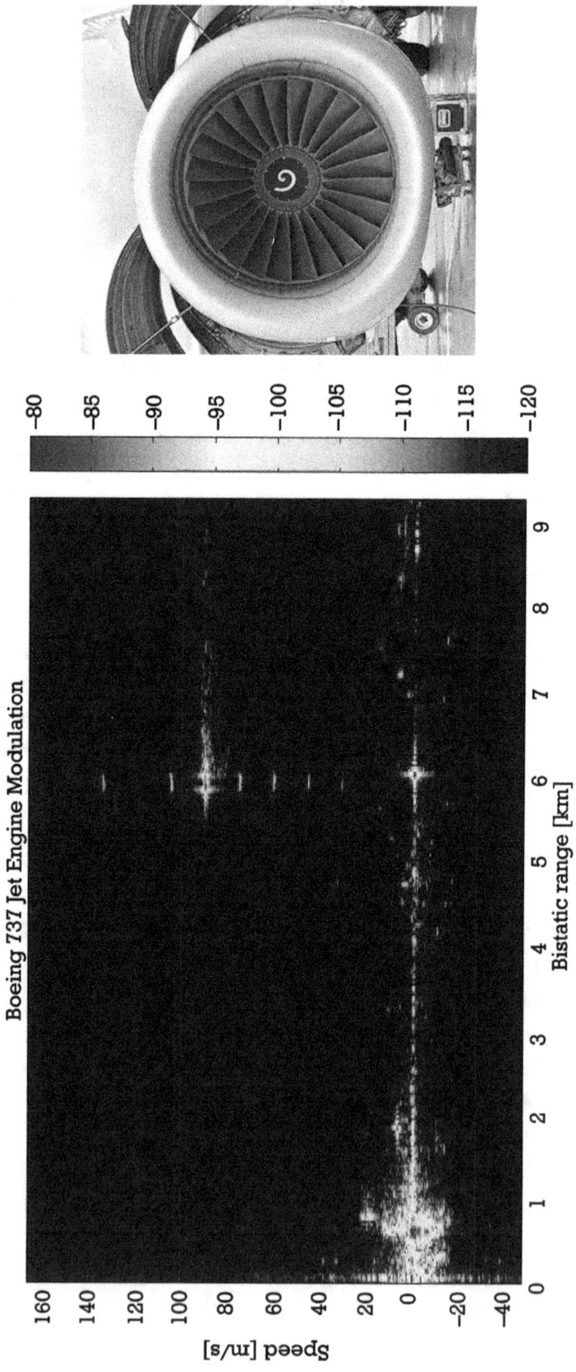

Figure 8.5 A range Doppler map of a Boeing 737 aircraft showing JEM lines after reflection from the jet engines [19].

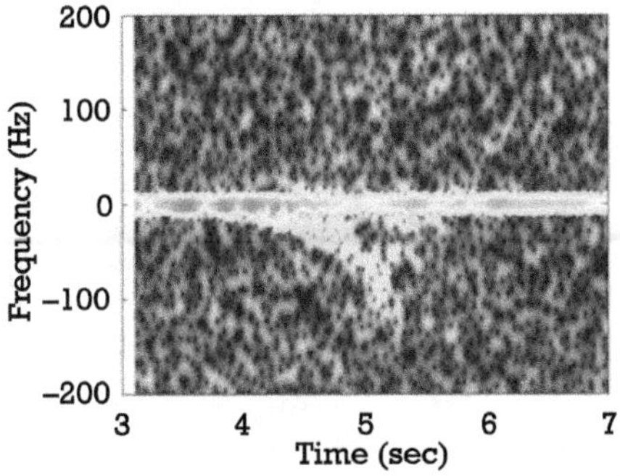

Figure 8.6 Spectrogram of a fall event obtained under laboratory conditions [23].

8.7 Low-Cost Passive Radar

One of the attractions of passive radar is that the receiver hardware can be simple and low-cost and, at the same time, achieve quite impressive performance.

Amateur radar enthusiasts have demonstrated some simple yet effective systems. As long ago as 1966, a publication in an amateur radio journal described some experiments to detect aircraft using a receiver located at Douai Abbey to the west of London in southern England and a VHF television transmitter at Lille in northeast France [24]. The geometry exploited the enhancement in target RCS in forward scatter, detecting the beat between the direct signal and the Doppler-shifted target echo. Figure 8.7 shows the path profile, taking into account the 4/3 effective Earth radius due to the fall in atmospheric refractive index with height, and also shows the regions where the target would be visible to both transmitter and receiver. The author also built a two-Yagi interferometer, such that a moving target would pass through the interferometer grating lobes, allowing the target motion to be estimated from the amplitude modulation.

Reference [25] described more recent amateur radio experiments, feeding the audio output from a high-frequency (HF) communications receiver to a personal computer, where it is digitized and processed by a simple fast Fourier transform (FFT) algorithm (downloadable from [25]), which presents the results as a spectrogram, an example of which is shown in Figure 8.8. This uses an HF transmitter at a frequency of about 26 MHz at a range of about 100 km, with the receiver rather closer to the targets. The spectrograms show the Doppler history of aircraft undertak-

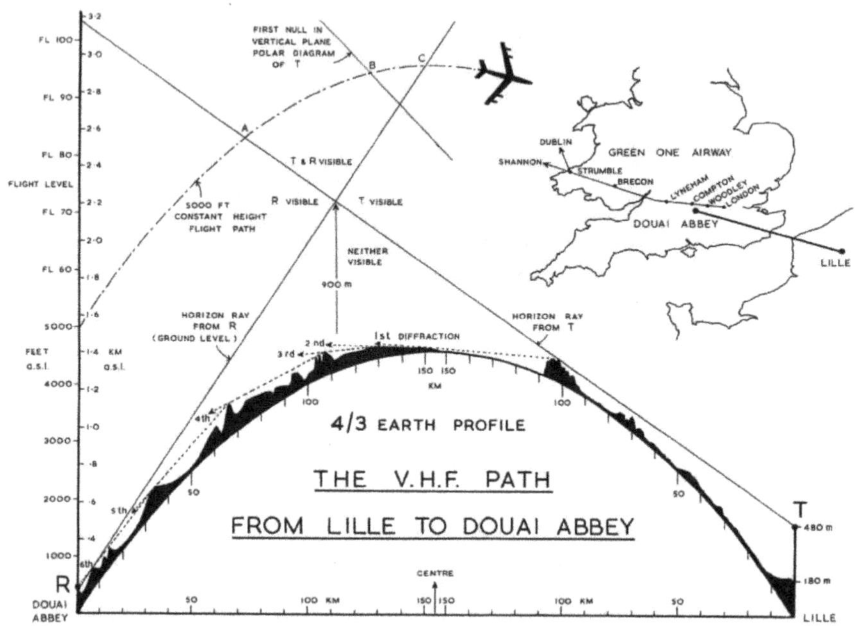

Figure 8.7 Profile of the path between the transmitter at Lille, northeast France, and the receiver at Douai Abbey, to the west of London [24]. (Copyright RSGB, used with permission.)

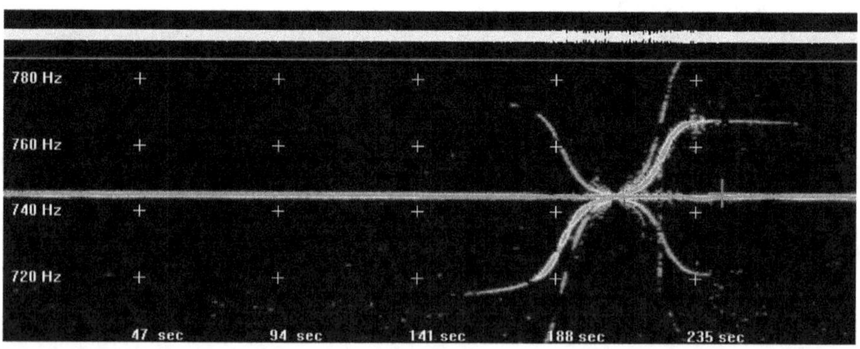

Figure 8.8 Spectrogram display of HF passive radar experiment. The vertical scale is Doppler shift in hertz, the horizontal scale is time in seconds. The central horizontal line is the direct signal carrier, and the features at around 200 seconds are due to echoes from an aircraft moving in a circular path [25].

ing various types of maneuver. As the web site [25] says, "Do try this at home." It is very easy.

Within research laboratories, the attraction of this approach was recognized as early as 1996, when Ogrodnik described and demonstrated a bistatic laptop radar [26]. The elements of a bistatic receiver system could be mounted in a single briefcase or carried in an aircraft.

More recently, the ready availability of "dongle" universal serial bus (USB) receivers means that a passive radar system can be assembled from little more than a laptop computer and a couple of antennas and software-defined radio USB devices. These are now available at low-cost (~$150) and with high performance. References [27, 28] described simple systems of this kind and the results achieved, including the use of a two-channel receiver to derive echo angle of arrival information.

These examples show what can be done with simple hardware and some ingenuity. It can be expected that, with even greater availability of sophisticated but low-cost hardware, the scope will increase still further.

8.8 The Intelligent Adaptive Radar Network

We assert that many of the conventional, monostatic approaches to military surveillance radar are inflexible, expensive, and vulnerable, and in future these functions will be better realized using intelligent, adaptive networks. The radars of the future will therefore be distributed, intelligent, spectrally efficient, and multistatic. A network of this kind is inherently resilient; if the nodes of the network are on moving platforms, such as UAVs, in the event of failure of one of the nodes the network can be reconfigured to restore its performance.

Passive radars may form part of such a network. Some nodes may be completely passive (receive-only), and if suitable broadcast, communications, or radionavigation signals are available, it makes sense to exploit them.

Challenges in realizing such a network include communication between the nodes of the network, especially if it is required to pass high-bandwidth raw data between platforms, geolocation and synchronization, especially in a GPS-denied environment, and overall management of the network. This latter problem has some similarities to that of resource management of a monostatic multifunction radar (MFR), but is manifestly more difficult. It is likely, however, that some of the techniques that have been developed for MFR resource management may be applied to this problem [29], as well as cognitive techniques which allow the radar to learn and adapt its operation [30].

8.9 Conclusions

It is interesting to recall Gordon Moore's visionary publication in 1965, just over 50 years ago [31], predicting essentially that computing power doubles every 18 months. By 2016, that measure processing power had increased by a factor of 1.7×10^{10} since the paper was published. By this measure, a calculation that today takes 1 millisecond would have taken 6.5 months in 1965. The very last word of Moore's paper is in fact "radar," showing that he understood the profound effect that his prediction would have on the capabilities of radar. There are physical limits to how far Moore's law can be extrapolated, but it can confidently be predicted that further increases in processing power will continue to have a big effect on what will be possible. In parallel, we can foresee advances in low-cost software-defined radio receivers and in sophisticated waveform coding and generation techniques.

The whole subject of spectrum allocation is set to undergo a revolution, both in regulation and in technology. Several functions that are now performed by conventional radars will in the future be realized by passive radar [32]. Advances can be expected in all of the topics mentioned in this chapter and in those of Chapter 7. In a military context, the desire for stealth will imply greater use of passive techniques. Passive radar is also an attractive technique for the monitoring of national land and maritime borders.

It would be wrong to suppose that passive techniques will overtake conventional radar technology. But with tongue only slightly in cheek, we can say, "The future is bright. The future is passive."

References

[1] Griffiths, H. D., et al., "Radar Spectrum Engineering and Management: Technical and Regulatory Approaches," *IEEE Proceedings*, Vol. 103, No. 1, January 2015, pp. 85–102.

[2] McQueen, D., "The Momentum Behind LTE Adoption," *IEEE Communications Magazine*, Vol. 47, No. 2, February 2009, pp. 44–45.

[3] Marcus, M. J., "Spectrum Policy for Radio Spectrum Access," *IEEE Proceedings*, Vol. 100, No. 5, May 2012, pp. 685–691.

[4] Baylis, C., et al., "Designing Transmitters for Spectral Conformity: Power Amplifier Design Issues and Strategies," *IET Radar, Sonar & Navigation*, Vol. 5, No. 6, July 2011, pp. 681–685.

[5] Oswald, G. K. A., "Holographic Radar," *Proceedings of SPIE 7308, Radar Sensor Technology XIII*, Orlando FL, April 2009.

[6] Sturm, C., and W. Wiesbeck, "Waveform Design and Signal Processing Aspects for Fusion of Wireless Communications and Radar Sensing," *IEEE Proceedings*, Vol. 99, No. 7, July 2011, pp. 1236–1259.

[7] Krier, J. R., et al., "Performance Bounds for an OFDM-Based Joint Radar and Communications System," *IEEE MILCOM 2015*, Tampa FL, October 26–28, 2015, pp. 511–516.

[8] Blunt, S. D., P. Yatham, and J. Stiles, "Intrapulse Radar-Embedded Communications," *IEEE Trans. on Aerospace and Electronic Systems*, Vol. 46, No. 3, July 2010, pp. 1185–1200.

[9] Griffiths, H. D., I. Darwazeh, and M. R. Inggs, "Waveform Design for Commensal Radar," *IEEE Int. Conference RADAR 2015*, Arlington VA, May 11–14, 2015, pp. 1456–1460.

[10] Stevens, M., D. Pompairac, and N. Millet, "Multi-Static Primary Surveillance Radar assessment," *SEE Int. Radar Conference RADAR 2014*, Lille, France, October 13–17, 2014.

[11] *Air Scientific Intelligence Interim Report, Heidelberg*, A.D.I. (Science), IIE/79/22, 24 Public Records Office, Kew, London (AIR 40/3036), November 24, 1944.

[12] Griffiths, H. D., "Klein Heidelberg: New Information and Insight," *IEEE Radar Conference 2015*, Johannesburg, October 27–30, 2015.

[13] Griffiths, H. D., "The D-Day Deception Operations TAXABLE and GLIMMER," *IEEE AES Magazine*, Vol. 30, No. 3, March 2015, pp. 12–20.

[14] Schüpbach, C., and U. Böniger, "Jamming of DAB-Based Passive Radar Systems," *EuRAD Conference 2016*, London, October 6–7, 2016.

[15] Griffiths, H. D., et al., "Denial of Bistatic Hosting by Spatial-Temporal Waveform Design," *IEE Proc. Radar, Sonar and Navigation*, Vol. 152, No. 2, April 2005, pp. 81–88.

[16] Pisane, J., et al., "Automatic Target Recognition (ATR) for Passive Radar," *IEEE Trans. on Aerospace and Electronic Systems*, Vol. 50, No. 1, January 2014, pp. 371–392.

[17] Ehrman, L. M., and A. Lanterman, "Automated Target Recognition Using Passive Radar and Coordinated Flight Models," *Proc. SPIE 5094, Automatic Target Recognition XIII*, Vol. 196, September 2003.

[18] Olivadese, D., et al., "Passive ISAR with DVB-T Signals," *IEEE Trans. on Geoscience and Remote Sensing*, Vol. 51, No. 8, August 2013, pp. 4508–4517.

[19] Garry, L., "Multistatic Passive Radar ISAR Imaging," PhD thesis, Ohio State University, Columbus, OH, 2016.

[20] Blacknell, D., and H. D. Griffiths, (eds.), *Radar Automatic Target Recognition and Non-Cooperative Target Recognition*, IET, Stevenage, August 2013.

[21] Ahmad, F., R. Narayanan, and D. Schreurs, "Application of Radar to Remote Patient Monitoring and Eldercare," *IET Radar, Sonar and Navigation*, Vol. 9, No. 2, February 2015, p. 115.

[22] Liang, L., et al., "Automatic Fall Detection Based on Doppler Radar Motion Signature," *5th International Conference on Pervasive Computing Technologies for Healthcare (PervasiveHealth)*, Dublin, May 23–26, 2011, pp. 222–225.

[23] Qisong, W., et al., "Radar-Based Fall Detection Based on Doppler Time-Frequency Signatures for Assisted Living," *IET Radar, Sonar and Navigation*, Vol. 9, No. 2, February 2015, pp. 164–172.

[24] Sollom, P. W., "A Little Flutter on VHF," *RSGB Bulletin*, November 1966, pp. 709–728; December 1966, pp. 794–824, www.rsgb.org.

[25] http://www.qsl.net/g3cwi/doppler.htm, accessed August 11, 2016.

[26] Ogrodnik, R. F., "Bistatic Laptop Radar: An Affordable, Silent Radar Alternative," *IEEE Radar Conference*, Ann Arbor MI, May 13–16, 1996, pp. 369–373.

[27] http://www.rtl-sdr.com/building-a-passive-radar-system-with-an-rtl-sdr/, accessed August 11, 2016.

[28] http://hackaday.com/2015/06/05/building-your-own-sdr-based-passive-radar-on-a-shoestring/, accessed August 11, 2016.

[29] Charlish, A., K. Woodbridge, and H. D. Griffiths, "Phased Array Radar Resource Management Using Continuous Double Auction," *IEEE Trans. on Aerospace and Electronic Systems*, Vol. 51, No. 3, July 2015, pp. 2212–2224.

[30] Haykin, S., "Cognitive Radar: A Way of the Future," *IEEE Signal Processing Magazine*, Vol. 23, No. 1, January 2006, pp. 30–40.

[31] Moore, G. E., "Cramming More Components onto Integrated Circuits," *Electronics*, April 19, 1965, pp. 114–117; reprinted in *IEEE Proceedings*, Vol. 86, No. 1, January 1998, pp. 82–85.

[32] Kuschel, H., and K. E. Olsen, (eds.), Special Issues of *IEEE AES Magazine* on Passive Radars for Civilian Applications, February 2017 and April 2017.

Bibliography

The following list identifies key papers, books and book chapters, and special issues of journals on bistatic radar in general and passive radar in particular.

Cherniakov, M., (ed.), *Bistatic Radar: Emerging Technology,* New York: Wiley, 2008.

Cherniakov, M., (ed.), *Bistatic Radar: Principles and Practice,* New York: Wiley, 2007.

Chernyak, V. S., *Fundamentals of Multisite Radar Systems: Multistatic Radars and Multiradar Systems,* Amsterdam: Gordon & Breach, 1998.

Dunsmore, M. R. B., "Bistatic Radars," Ch. 11 in *Advanced Radar Techniques and Systems,* G. Galati, (ed.), London. U.K.: Peter Peregrinus, 1993.

Farina, A., and H. Kuschel, (eds.), Special Issues of *IEEE AES Magazine on Passive Radar,* Vol. 27, No. 10, October 2012.

Farina, A., and H. Kuschel, (eds.), Special Issues of *IEEE AES Magazine on Passive Radar,* Vol. 27, No. 11, November 2012.

Glaser, J. I., "Fifty Years of Bistatic and Multistatic Radar," *IEE Proc.,* Vol. 133, Pt. F, No. 7, December 1986, pp. 596–603.

Griffiths, H. D., "From a Different Perspective: Principles, Practice and Potential of Bistatic Radar," *Proc. Int. Radar Conference RADAR 2003,* Adelaide, Australia, September 3–5, 2003, pp. 1–7.

Griffiths, H. D., "New Directions in Bistatic Radar," *IEEE Radar Conference 2008*, Rome, Italy, May 26–29, 2008.

Griffiths, H. D., "Workshop on Multistatic and MIMO Radar," *IEEE AES Magazine*, Vol. 25, No. 2, February 2010, pp. 43–45.

Griffiths, H. D., and C. J. Baker, "Passive Bistatic Radar," Ch. 11 in *Principles of Modern Radar: Radar Applications*, W. Melvin and J. Scheer, (eds.), Raleigh, NC: SciTech Publishing, 2014.

Griffiths, H. D., and N. R. W. Long, "Television-Based Bistatic Radar," *IEE Proc.*, Vol. 133, Pt. F, No. 7, December 1986, pp. 649–657.

Griffiths, H. D., and C. Stewart, "Vectors in Radar Technology," *IEEE AES Magazine*, Vol. 31, No. 8, August 2016, pp. 41–45.

Howland, P. E., (ed.), Special Issue of *IEE Proceedings on Radar, Sonar & Navigation on Passive Radar Systems*, Vol. 152, No. 3, June 2005.

Howland, P. E., "Target Tracking Using Television-Based Bistatic Radar," *IEE Proc. Radar, Sonar and Navigation*, Vol. 146, No. 3, June 1999, pp. 166–174.

Howland, P. E., H. D. Griffiths, and C. J. Baker, "Passive Bistatic Radar," in *Bistatic Radar: Emerging Technology*, M. Cherniakov, (ed.), New York: Wiley, 2008.

Howland, P. E., D. Maksimiuk, and G. Reitsma, "FM Radio Based Bistatic Radar," *IEE Proc. Radar, Sonar and Navigation*, Vol. 152, No. 3, June 2005, pp. 107–115.

Jackson, M. C., "The Geometry of Bistatic Radar Systems," *IEE Proc.*, Vol. 133, Pt. F, No. 7, December 1986, pp. 604–612.

Kell, R. E., "On the Derivation of Bistatic RCS from Monostatic Measurements," *Proc. IEEE*, Vol. 53, August 1965, pp. 983–988.

Kuschel, H., and K. E. Olsen, (eds.), Special Issue of *IEEE AES Magazine on Passive Radars for Civilian Applications*, February 2017.

Kuschel, H., and K. E. Olsen, (eds.), Special Issue of *IEEE AES Magazine on Passive Radars for Civilian Applications*, April 2017.

Lombardo, P., and F. Colone, "Advanced Processing Methods for Passive Bistatic Radar Systems," Ch. 17 in *Principles of Modern Radar: Advanced Techniques*, W. Melvin and J. Scheer, (eds.), Raleigh, NC: SciTech Publishing, 2013.

Pell, C., and E. Hanle, (eds.), Special issue of *IEE Proceedings Part F on Bistatic Radar, IEE Proc.*, Vol. 133, Pt.F, No. 7, December 1986.

Skolnik, M. I., "An Analysis of Bistatic Radar," *IEEE Trans. Aerospace and Navigational Electronics,* March 1961, pp. 19–27.

Tsao, T., et al., "Ambiguity Function for Bistatic Radar," *IEEE Trans. on Aerospace and Electronic Systems,* Vol. AES-33, 1997, pp. 1041–1051.

Willis, N. J., *Bistatic Radar,* 2nd ed., Silver Spring, MD: Technology Service Corp., 1995; corrected and republished, Raleigh NC: SciTech Publishing, 2005.

Willis, N. J., "Bistatic Radar," Ch. 23 in *Radar Handbook,* 3rd ed., M. I. Skolnik, (ed.), New York: McGraw-Hill, 2008.

Willis, N. J., and H. D. Griffiths, (eds.), *Advances in Bistatic Radar,* Raleigh, NC: SciTech Publishing, 2007.

About the Authors

Hugh Griffiths holds the THALES/Royal Academy Chair of RF Sensors at University College London, United Kingdom. He has published over 500 papers in the fields of radar, antennas, and sonar. In 1996, he received the IEEE AES Nathanson Award, and in 2012 he was awarded the IET A.F. Harvey Prize for his work on bistatic radar. He has also received the Brabazon Premium of the IERE, the Mountbatten and Maxwell Premium Awards of the IEE, and the 2015 IEEE AES Mimno Award. He is a Fellow of the IET and a Fellow of the IEEE, and in 1997 he was elected to Fellowship of the Royal Academy of Engineering. He served as the president of the IEEE Aerospace and Electronic Systems Society for 2012 to 2013.

Chris Baker is the chief technology officer with Aveillant Ltd. in Cambridge, United Kingdom, where he works on commercialization of Aveillant's 3D Holographic Radar technology. Previously, he was the Ohio Research Scholar in Integrated Sensor Systems at The Ohio State University, United States. He has received the IEE Mountbatten Premium (twice) and the IEE Institution Premium. He is a Fellow of the IEEE and of the IET. His research interests include coherent radar techniques, radar signal processing, radar signal interpretation, electronically scanned radar systems, natural echo locating systems, and radar imaging, and he is the author of more than 250 publications.

Index

Recent Titles in the Artech House Radar Series

Dr. Joseph R. Guerci, Series Editor

Adaptive Antennas and Phased Arrays for Radar and Communications, Alan J. Fenn

Advanced Techniques for Digital Receivers, Phillip E. Pace

Advances in Direction-of-Arrival Estimation, Sathish Chandran, editor

Airborne Pulsed Doppler Radar, Second Edition, Guy V. Morris and Linda Harkness, editors

Basic Radar Analysis, Mervin C. Budge, Jr. and Shawn R. German

Bayesian Multiple Target Tracking, Second Edition Lawrence D. Stone, Roy L. Streit, Thomas L. Corwin, and Kristine L Bell

Beyond the Kalman Filter: Particle Filters for Tracking Applications, Branko Ristic, Sanjeev Arulampalam, and Neil Gordon

Cognitive Radar: The Knowledge-Aided Fully Adaptive Approach, Joseph R. Guerci

Computer Simulation of Aerial Target Radar Scattering, Recognition, Detection, and Tracking, Yakov D. Shirman, editor

Control Engineering in Development Projects, Olis Rubin

Design and Analysis of Modern Tracking Systems, Samuel Blackman and Robert Popoli

Detecting and Classifying Low Probability of Intercept Radar, Second Edition, Phillip E. Pace

Digital Techniques for Wideband Receivers, Second Edition, James Tsui

Electronic Intelligence: The Analysis of Radar Signals, Second Edition, Richard G. Wiley

Electronic Warfare in the Information Age, D. Curtis Schleher

Electronic Warfare Target Location Methods, Second Edition, Richard A. Poisel

ELINT: The Interception and Analysis of Radar Signals, Richard G. Wiley

EW 101: A First Course in Electronic Warfare, David Adamy

EW 102: A Second Course in Electronic Warfare, David Adamy

EW 103: Tactical Battlefield Communications Electronic Warfare, David Adamy

Fourier Transforms in Radar and Signal Processing, Second Edition, David Brandwood

Fundamentals of Electronic Warfare, Sergei A. Vakin, Lev N. Shustov, and Robert H. Dunwell

Fundamentals of Short-Range FM Radar, Igor V. Komarov and Sergey M. Smolskiy

Handbook of Computer Simulation in Radio Engineering, Communications, and Radar, Sergey A. Leonov and Alexander I. Leonov

High-Resolution Radar, Second Edition, Donald R. Wehner

Highly Integrated Low-Power Radars, Sergio Saponara, Maria Greco, Egidio Ragonese, Giuseppe Palmisano, and Bruno Neri

Introduction to Electronic Defense Systems, Second Edition, Filippo Neri

Introduction to Electronic Warfare, D. Curtis Schleher

Introduction to Electronic Warfare Modeling and Simulation, David L. Adamy

Introduction to RF Equipment and System Design, Pekka Eskelinen

Introduction to Modern EW Systems, Andrea De Martino

An Introduction to Passive Radar, Hugh D. Griffiths and Christopher J. Baker

The Micro-Doppler Effect in Radar, Victor C. Chen

Microwave Radar: Imaging and Advanced Concepts, Roger J. Sullivan

Millimeter-Wave Radar Targets and Clutter, Gennadiy P. Kulemin

Modern Radar Systems, Second Edition, Hamish Meikle

Modern Radar System Analysis, David K. Barton

Modern Radar System Analysis Software and User's Manual, Version 3.0, David K. Barton

Monopulse Principles and Techniques, Second Edition, Samuel M. Sherman and David K. Barton

MTI and Pulsed Doppler Radar with MATLAB®, Second Edition, D. Curtis Schleher

Multitarget-Multisensor Tracking: Applications and Advances Volume III, Yaakov Bar-Shalom and William Dale Blair, editors

Precision FMCW Short-Range Radar for Industrial Applications, Boris A. Atayants, Viacheslav M. Davydochkin, Victor V. Ezerskiy, Valery S. Parshin, and Sergey M. Smolskiy

Principles of High-Resolution Radar, August W. Rihaczek

Principles of Radar and Sonar Signal Processing, François Le Chevalier

Radar Cross Section, Second Edition, Eugene F. Knott, et al.

Radar Equations for Modern Radar, David K. Barton

Radar Evaluation Handbook, David K. Barton, et al.

Radar Meteorology, Henri Sauvageot

Radar Reflectivity of Land and Sea, Third Edition, Maurice W. Long

Radar Resolution and Complex-Image Analysis, August W. Rihaczek and Stephen J. Hershkowitz

Radar RF Circuit Design, Nickolas Kingsley and J. R. Guerci

Radar Signal Processing and Adaptive Systems, Ramon Nitzberg

Radar System Analysis, Design, and Simulation, Eyung W. Kang

Radar System Analysis and Modeling, David K. Barton

Radar System Performance Modeling, Second Edition, G. Richard Curry

Radar Technology Encyclopedia, David K. Barton and Sergey A. Leonov, editors

Radio Wave Propagation Fundamentals, Artem Saakian

Range-Doppler Radar Imaging and Motion Compensation, Jae Sok Son, et al.

Robotic Navigation and Mapping with Radar, Martin Adams, John Mullane, Ebi Jose, and Ba-Ngu Vo

Signal Detection and Estimation, Second Edition, Mourad Barkat

Signal Processing in Noise Waveform Radar, Krzysztof Kulpa

Space-Time Adaptive Processing for Radar, Second Edition, Joseph R. Guerci

Special Design Topics in Digital Wideband Receivers, James Tsui

Theory and Practice of Radar Target Identification, August W. Rihaczek and Stephen J. Hershkowitz

Time-Frequency Signal Analysis with Applications, Ljubiša Stanković, Miloš Daković, and Thayananthan Thayaparan

Time-Frequency Transforms for Radar Imaging and Signal Analysis, Victor C. Chen and Hao Ling

Transmit Receive Modules for Radar and Communication Systems, Rick Sturdivant and Mike Harris

For further information on these and other Artech House titles, including previously considered out-of-print books now available through our In-Print-Forever® (IPF®) program, contact:

Artech House
685 Canton Street
Norwood, MA 02062
Phone: 781-769-9750
Fax: 781-769-6334
e-mail: artech@artechhouse.com

Artech House
16 Sussex Street
London SW1V HRW UK
Phone: +44 (0)20 7596-8750
Fax: +44 (0)20 7630-0166
e-mail: artech-uk@artechhouse.com

Find us on the World Wide Web at: www.artechhouse.com

www.ingramcontent.com/pod-product-compliance
Lightning Source LLC
LaVergne TN
LVHW051050220926
848485LV00001BA/62